SUM AND SUBSTANCE

Comprehending Yourself,
Your Universe, and Your World

William L. Martin

Auburn, CA
Copyright © 2022 William L. Martin
First edition
This book is copyright under the Berne Convention
All rights reserved. No reproduction without permission.

ISBN: 979-8-9860657-0-0
Library of Congress Control Number: 2022907123

Printed in the United States of America

For Linda Lupher Martin

"My universe, my world, my life"

Preface

———⊶⊷⊶———

My riches are these poor habiliments,
Of which if you should here disfurnish me,
You take the sum and substance that I have.
William Shakespeare
(1564 – 1616)

Life can only be understood backwards,
but it must be lived forward.
Soren Kierkegaard
(1813 – 1855)

This book has the ambitious hope that, if you spend eight hours reading it, you may learn more than in any other eight hours of your life.

Most knowledge about the universe and humanity's place in it has been learned in the last four hundred years, as a result of science, technology, and education. This work summarizes the path of humanity in terms understandable to anyone who will try. "Sum and substance," as Shakespeare said, but little else.

The book has only two prerequisites: You must be able to read, and you must have some grasp of numbers, large and small. The first I can assume as you have made it this far. The second is less certain, so I include a tutorial (Chapter 3) to help. Even if you are very proficient with numbers, you might find something there to enjoy.

All dates follow the modern convention in science that CE denotes "Common Era" which is the same as AD, "Anno Domini," and BCE denotes "Before Common Era" which is the same as BC, "Before Christ." Dates having no suffix are implicitly CE. Temperatures are given either in the Centigrade scale (water freezing at 0 degrees C.) or the Fahrenheit scale (water freezing at 32 degrees F.)

I intentionally violate two principles of scholarly writing. First, I tell WHAT has been learned with little attribution to WHO first learned it. The goal is to convey as much information as possible in few words, to keep this work more focused. Historical context and credit to key contributors are important, but not here, and I offer no apology for failing to give credit where due. Credit is implied with humble gratitude and admiration, but Isaac Newton, Albert Einstein, and others long deceased do not care that I occasionally omit their names. Second, you will find no footnotes or references for similar reasons. I have read hundreds of sources in preparing this work, and to list them would add nothing of value. A bibliography of key readings is provided should you care to go further in any topic.

In seeking sum and substance, many topics are covered without much depth. Results which have required thousands of lifetimes to achieve are treated in a sentence or two. Some normally important topics are omitted altogether. Also, every discipline has debates among scholars over the details. You may wish to further investigate any that interest you, but be aware that a lifetime of study might be required.

This book summarizes the entire body of work of humanity as seen through one pair of eyes (mine). Nobody else would see it in exactly the same way. I claim nothing original, and you are welcome to disagree with anything said. If we disagree, please know that I intend you no offense. I have written this in my ninth decade, and it represents my understanding of life gained backwards. "My riches are these poor habiliments." My hope is to add something to your life as you see it through your own eyes, understanding backwards, but living forward.

Take this journey to a richer comprehension of your life and its meaning, the evolution of our universe as understood by science, and the path of civilizations from the first humans to modern times.

Table of Contents

Preface   v

PART 1 – COMPREHENDING YOURSELF

Chapter 1: Allow Me to Introduce Yourself –
            Comprehension Begins at Home   3

Chapter 2: The Scientific Method –
            How We Know Things   20

Chapter 3: Numbers Large and Small –
            Learning How to Count   30

Chapter 4: Roadmap – Milestones of Your Universe
            and Your World   41

PART 2 – COMPREHENDING YOUR UNIVERSE

Chapter 5: The Universe – BANG! You Are Alive   64

Chapter 6: The Milky Way – A Galaxy of Stars   75

Chapter 7: The Sun – The Light of Your Life   83

Chapter 8: Earth – Your Home   91

Chapter 9: Life on Earth – From One Cell to the Next   101

Chapter 10: Homo Sapiens – The Organism   112

PART 3 – COMPREHENDING YOUR WORLD

Chapter 11: Preview of Part 3   124

Chapter 12: Hunter/Gatherers – Down From the Trees   127

Chapter 13: Civilizations – From the Farm to the Empire   139

Chapter 14: Feudalism to Domination –
            Europe Conquers the World   153

Chapter 15: Modern Times – Revolutions
            and Ideologies   166

Chapter 16: Warfare – Bang! You Are Dead   188

Chapter 17: Systems of Belief –
    Everyone is in the Minority   202

Chapter 18: The Biosphere – To Be or Not to Be   216

PART 4 – FROM COMPREHENSION TO CONJECTURE AND CONCLUSION

Chapter 19: The Meaning of Life –
    It Is Your Choice   232

Chapter 20: The Future – Where There is Life,
    There is Hope (And Fear)   241

Chapter 21: Ideas – From the Trivial to the Colossal   253

Chapter 22: A Summary in Quotations –
    Your Conversation with the Ages   262

Chapter 23: Conclusion – Summary of the
    Sum and Substance   275

Bibliography   291

Acknowledgements   294

About the Author   295

Part 1
Comprehending Yourself

Chapter 1
Allow Me to Introduce Yourself – Comprehension Begins at Home

―――∞∞∞―――

Always remember that you are absolutely unique.
Just like everyone else.
Margaret Mead
(1901 – 1978)

So, remember when you're feeling very small and insecure
How amazingly unlikely is your birth.
Monty Python
(1969 – 2014)

Congratulations, dear reader! You are the result of four billion years of development of life on Earth. You are amazing in at least four distinct ways. First, you overcame nearly insurmountable odds to come into existence. Second, you had the good fortune to be born as a member of the species Homo sapiens, with mental capabilities far in excess of the millions of other species populating our planet. Third, you live in the twenty-first century and are the beneficiary of the explosion of science and technology over the last four hundred years. Fourth, you are privileged to experience this beautiful planet together with the wonderful accomplishments of those who came before you – Beethoven, Michelangelo, Einstein, Monty Python, and your mother. You can know music, you can know beauty, you can know science, you can know humor, and you can know love. You are a wonder!

Alas, there are also some downsides to life, but we will put those aside for now, returning to them later. For the moment, please rejoice in the positive.

Your personal journey began two or three decades before your birth. When your mother was born, she began her life with about a million eggs. By puberty, she had perhaps 300,000 remaining and ovulated only about three or four hundred. One of those eggs had half of your DNA. Somehow, she encountered your father. If he was a normal male, his sperm count was perhaps 40 million per dose. One of those containing the other half of your DNA won the race to your mom's egg, and they united to form you and your complete DNA. So, at birth, you beat odds of forty quadrillion to one. That is longer odds than any lottery you will ever enter!

Of course, your debt to your parents is greater than your conception. Your mom decided to give birth notwithstanding that it was one of the riskiest events of her life. Then she was confronted with this helpless creature (you) who demanded enormous time and energy for decades. Hopefully she and your dad loved you, and you were more joy than burden.

Although you did not choose your parents, they determined much about your future prospects. They each supplied half of your DNA, which specifies much about your prospects for good health, innate intelligence, and every other biological characteristic of your mind and body. Your parents determined your race and sex. You lived where they lived, whether on Manhattan Island or in Hong Kong. You speak their language, from among the 7,000 living languages. (More people on Earth speak Mandarin Chinese than any other, over a billion, with Hindi, Spanish, and English having about half a billion each.)

If your parents were well off financially, probably you were too. Hopefully, you were not among the 700 million or so who are undernourished. The world is doing much better in this regard now than at any other time in history. In fact, there are about as many people in the world who are obese, which is itself a substantial health problem. Obesity is currently growing as a major health issue at a faster rate than malnourishment.

Who you became is largely determined by your environment in the first decade of your life. Let me tell you a little about my early life, because it might help revive memories of your own childhood.

4

My parents married in Pennsylvania in 1932 during the Great Depression, which influenced the rest of their lives. I was their third child, born in 1939 in the same week that Albert Einstein informed President Roosevelt that an atomic weapon could be produced. Hitler had just begun his program of extermination of Polish Jews. Our family moved to Phoenix, Arizona, in 1942, a few months after Pearl Harbor, in search of a better, post-depression life. I was blissfully unaware, except I recall playing a game called "step on a crack, and break Hitler's back."

My dad regaled me with stories of sneaking into Forbes Field in Pittsburgh to watch his beloved Pirates play baseball. He played football at Duquesne University and carried a sore shoulder for life as a result. He was an accomplished violinist who performed with the Pittsburgh Symphony Orchestra. He trained himself to be one of the earliest computer programmers. It is no coincidence that I am an avid sports fan and have a subscription to the Digital Symphony Hall of the Berlin Philharmonic, even though I failed to inherit his musical talent. I spent my high school years as a gym rat playing basketball, and injured a shoulder which required surgery, and I chose to be a computer hardware designer.

My mom was a registered nurse who took a low-key approach to parenting. The opposite of the contemporary "helicopter" mom, she described her approach to parenting as "just let them grow." When I went out to play with friends in the morning, the only guidance was "be home for dinner." There was no sense of danger in the community. Her reaction to trouble was always "no use fussing."

I joke that my early moral training came from comic books and radio shows. The cell phone, game box, and television were far in the future. My favorite heroes were Superman, who promoted the virtues of "truth, justice and the American way," and the Lone Ranger. Every radio episode of the latter began with these immortal words:

"With his faithful Indian companion, Tonto, the daring and resourceful Masked Rider of the Plains led the fight for law and order in the early western United States. Nowhere in the pages of history can one find a greater champion of justice! Return with us now to those thrilling days of yesteryear! From out of the past come the thundering hoofbeats of the great horse, Silver. The Lone Ranger rides again!"

This soliloquy was accompanied by a rousing rendition of the William Tell Overture. One definition of an intellectual was "a person who can hear the William Tell Overture without thinking of the Lone Ranger."

In both comics and radio shows, the boundaries between good and evil were clear. The Lone Ranger always got his man, but he never shot to kill. Superman saved Lois Lane, but nobody really got hurt. I did not know that "Tonto" translates to "stupid" in Spanish, and it never occurred to me that Tonto's grunting responses were a blatant racial stereotype.

I lived in a home where race was never mentioned. It did not occur to me until many years later that I had grown up in a city with strict segregation. I knew that there were black people, but they all lived "south of the tracks," and we never went there. Because I was a fan of high school football and basketball, I knew that there was a local high school called Carver High located on the other side of town. The team I rooted for, North Phoenix High, was not allowed to play them.

The main point of this monolog about my memories is that every single word emerged from my mind. We define ourselves almost completely by what is stored in our brains. When people analyze what makes humans unique, three main attributes are usually named. First, we are bipedal. That is, we walk on two legs rather than four. This frees up our arms to harvest wheat and throw spears. Second, we have opposable thumbs, which allows us to use tools and draw pictures. But by a wide margin, our brains are what enable our species to seek to control our environment rather than merely reside in it.

Imagine for a moment that your brain is portable, and you could place it on a table. You would find that it weighs about two percent of your body weight and looks like highly wrinkled tofu. Its volume is a little more than a quart, and it is about 75 percent water.

In spite of being only two percent of your weight, the brain consumes about 20 percent of your energy, about 20 percent of your oxygen intake, and about 60 percent of your blood sugar. About 25 percent of the body's cholesterol is in the brain.

It is so costly because it is busy. The brain controls most of the involuntary functions of the body, including breathing and blinking your eyes. It processes information about all five of your senses – sight, hearing, touch, smell, and taste. It integrates whatever presents itself to you and tries to make sense of it. It enables your uniquely human skills – reading, writing, creating stories, composing music, betting on the Super Bowl, contemplating the stars, and comprehending your place in the universe.

The brain accomplishes this via what is often regarded as the most complex structure ever identified. A neuron is the fundamental unit of the brain that receives input from the world and relays the interpretation throughout your body. You have about 100 billion of them. A synapse is the junction that interconnects the neurons. Each neuron connects to about 40 thousand synapses. Altogether, it has been estimated your brain has a total capacity of about a quadrillion bytes, which is more than 20 billion times the total number of words in the English language. (See Chapter 3 for further insight into numbers this large.)

In spite of all research done on the brain, there is no general consensus as to how brain capacity turns into a living, conscious human being. The physical brain and the conscious mind, and the resulting person, are not identical. What is clear is that for many reasons, starting with our parents' teachings, we all experience reality individually. There is one Sun shining down on us, but all 7.8 billion of us perceive the world in our own unique way and reach differing viewpoints. Perhaps the beginning of charity is to recognize that no one else experiences reality in exactly the same way you do.

As you are indebted to your parents, so were they indebted to theirs. Let us envision your entire family tree of direct ancestors. You have two parents, four grandparents, eight great-grandparents, sixteen great-great grandparents and so forth, with the number doubling at each generation. This doubling continues as far into the past as you can imagine. Each was a real person who led a real life, and if any one had failed to exist, there would be no "you."

Somewhere along the line, there were probably close calls. For example, my wife's great grandfather, Jacob, and his wife, Gertrude, conceived her grandfather shortly before Jacob was killed at the battle of Pea Ridge in the American Civil War. If Jacob had died a few months earlier, then my wife and my children would not exist.

The number of your ancestors doubles with each generation, but going back in time, the world population was significantly less than today. Let us do the exercise of determining how far back in time we have to go until the number of your ancestors exceeds the human population. I think the result may surprise you.

We can take 25 years as the average length of a generation. For example, four generations ago, in about 1921, you had eight great-great-great grandparents with this number doubling every 25 years. In 1921, the population of the world was a little more than a billion. If we continue to make this comparison each 25 years, we discover that in the 29th generation, in the year 1296, you have 536 million direct ancestors and the population of the world was about 400 million. In other words, back a surprisingly short period of time, you are related to just about everybody in the world! If you are of European descent, then the odds are high that you are related directly to almost everyone who signed the Magna Carta in 1225. There was a Guilliam Martin who fought for William the Conqueror in 1066, and the chances are good that they are both my direct ancestors.

Actually, the analysis is not quite this simple since many human groups were disconnected by migration from each other for long periods. So, if your ancestral roots are in Asia, you might not have a common ancestor with a German as late as the 13th century. However, going back even further in time, before humans migrated out of Africa, the more certain it becomes that you are related to everyone.

Let us continue our thought experiment of going back a generation at a time in 25-year increments, except this time imagine that we go back thousands and then millions and then a few billion years. If you go back 250 thousand years, your ancestors are still human beings, and you could reproduce children with them. A million years ago, your ancestors would still be recognizable hominids, but you could not have children together. You would no longer be the same species. Finally, you go back further in time, in the unbroken chain of life, reaching the instant in time when the first single-cell life appeared on Earth. Everything alive on Earth shares the common bond of DNA as the fundamental means of reproduction and continuity. Not only do you share over 99 percent of your DNA with a chimpanzee, you share 50 percent with a potato.

Having resolved who you are, let us see where you are. Perhaps, at the moment, you are sitting quietly in a comfortable place with good lighting. If I asked you where you live, you might give me a street address or the name of your town, state or nation. That would be true, but those locations are merely transitory political conveniences.

From a different perspective, consider your place on the planet and in space. You are not really sitting still; you are actually in motion in several different ways. First, if you are in the United States, your chair is sitting on the North American tectonic plate. Every point on Earth sits on one of seven such plates, and they are all in motion, albeit very slowly and punctuated by earthquakes and volcanic eruptions. The average speed of this motion is only about three to five centimeters per year. If you are currently in Southern California, you will be relocated to where San Francisco now sits in about eight million years – a short interval in the sands of time.

While your tectonic plate is moving northward at its leisurely pace, you are also spinning around the Earth's axis at the rate of once every 24 hours. The circumference of the Earth is about 24 thousand miles. Therefore if you live on the equator, you are spinning at the rate of about 1,000 miles per hour.

Also, you are clinging to a planet that is revolving around the Sun once every three hundred sixty-five and a quarter days, at a radius of 93 million miles. This means you are moving around the Sun at the rate of about 67 thousand miles per hour.

Further, you and the Sun and Earth are revolving around the center of the Milky Way galaxy, which has a radius of 53 thousand light years and rotates at the rate of once every 240 million years. (A light year is the distance that light travels in one year at the rate of 186 thousand miles per second.) This translates into your movement at the rate of about 475 thousand miles per hour. Since Homo sapiens first appeared about 250 thousand years ago, the Milky Way has rotated less than half a degree.

Finally, you, the Milky Way, the Sun, and Earth are part of a universe that is expanding at a rate which may approach the speed of light. And you thought you were just sitting quietly reading!

As recently as 500 years ago, Christendom was committed to the belief that Earth was stationary and everything else rotates around it. A principal argument was that if Earth were really rotating, then we would all be blown away by the wind. Of course, the fact is that everything on the planet, including its atmosphere, is rotating at the same rate so there is no relative motion caused by all the commotion of spinning, orbiting and expanding.

While we are discussing where you are, we can think about who owns the land you are sitting on. Perhaps you do. If you are renting, your landlord most likely holds title. If you are in a public place, like a library or school, the institution probably claims ownership. Of course, the property is governed by the applicable laws of city, county, state, and national governments.

These are all short-term arrangements. In the longer term, we are occupants as the beneficiaries of the last conqueror. For example, I live in the Mother Lode of California, site of one of the greatest migrations in history at the time of the Gold Rush in 1849. At the time, California was a territory of the United States and later became a state in 1850. Earlier, it had been a part of Mexico until 1821, when it was lost to the United States in the Mexican-American war. Prior to that, it was a part of the Spanish Empire from 1769 to 1821. Various European explorations reached the area beginning in 1542 and proceeded to destroy, via conquest and disease, most of the five Native American tribes who occupied this area for the previous ten thousand years.

This land was inhabited before humans arrived. It was the home where the buffalo roamed and where the deer and the antelope played. And sabertooth tigers. And dinosaurs. There were also eras when the area was under water, and even some where it was buried beneath glacier ice. No one knows for sure what the future will bring, except that a million years hence, the place where you are sitting will be very different than it is now.

We are lucky to be living in the twenty-first century. It may be of interest to compare your life now to what it might have been if you been born earlier. As examples, consider three years: 1821, 1921, and 2021. Imagine that in all three cases you lived on the

East Coast of the United States, and that you might have been rich or poor; male or female; white, black, or brown.

First, consider 1821. The population of the young United States was growing rapidly and had reached almost 10 million, of whom 1.2 million were slaves. Founding Fathers, Thomas Jefferson and John Adams, were still living and would remain so for another five years, before dying on the same day, July 4, 1826. Jefferson had owned about 400 slaves, including four to six he sired himself. George Washington had been gone for a couple of decades, having been bled to death by the medical expertise of the time. (Bleeding would be regarded as good medical practice for another sixty years.) James Monroe was in office as the fifth president, and would soon declare the Monroe Doctrine renouncing European colonialism in the Western Hemisphere. His presidency was called "The Era of Good Feelings," a feeling of national unity following the trauma of the war of 1812, when the British burned down the White House. Monroe's goal was to eliminate the partisanship inherent in party politics. Missouri became the 24th state and was admitted to the union as a slave state, frustrating northern attempts to stop slavery's expansion. The internal combustion engine, invented in 1821, was not widely in use. The first railroad in the United States would not be built for another six years. The main modes of transportation were walking, horseback, carriages, and sailing ships. The world record for circumnavigating the globe was 676 days, held since 1617 by Dutch navigator Willem Schouten. The Industrial Revolution in the United States was in its infancy, but textile mills had reached both New England and the South and were increasing in economic importance.

You could probably read and write since the United States had done a good job of making education compulsory. Only 12 percent of the people in the world could read and write, but the literacy rate in the United States was almost 90 percent among both non-slave men and women. The post office existed, so you could communicate remotely via mail. Often, letters were best carried by friends or merchants. Postal rates were extremely high, with a single sheet costing six cents to mail when the average labor rate was about 50 cents per day.

Women were not allowed to vote and usually could not own property in their own name. However, many states passed laws giving a woman the right to approve any sale of her home.

Ludwig von Beethoven was still alive in Germany and did not compose his ninth symphony for another six years. His music had first been performed in the United States in Charleston, South Carolina. At the time, Charleston was the center of the slave trade and home to many of the country's wealthiest men

Life expectancy in the United States in 1821 was about 41, held down by the high infant mortality rate. Women bore an average of six to seven children, but buried an average of three before their fifth birthday. If you lived to adulthood, your chances of a long life were good. Ben Franklin, George Washington, Thomas Jefferson, and John Hancock lived an average of 77 years, which is about the same as an American male can expect today. Betsy Ross, Dolley Madison, Martha Washington, and Martha Jefferson lived an average of 68 years, with the average being reduced by Martha Jefferson's premature death after childbirth when she was only 34.

By comparison, life expectancy six thousand years earlier, in ancient Sumeria, one of the earliest civilizations, has been estimated at between 35 and 40. Similar estimates apply to Ancient Rome and Athens, so that at least by this measure, humanity had not gained much.

In 1821, over 70 percent of the labor force worked as farmers having fallen rapidly from the 90 percent farm labor force of 1790, only thirty years earlier. Again, 90 percent is about the same percentage of farmers as in ancient Sumeria. Urbanization and the Industrial Revolution had begun to take effect by 1821, and these trends would accelerate rapidly now. Nevertheless, it is fair to say that, in many regards, life in 1821 America was more similar to life in ancient Sumeria than to life in 2021. If you wanted to talk to someone, you had to be in their presence. If you wanted to travel, you could walk, use a beast of burden (horse, ox or donkey), or sail on a boat. When the Sun went down, there was very little light. If you became ill, there was very little effective help for you. None of those things had changed much in 7,000 years.

Stepping forward a century, consider life in 1921. Population of the United States had grown tenfold to 106 million. The great Spanish flu pandemic of 1918 to 1920 was ending, having killed between 20 and 50 million people worldwide including almost 700 thousand Americans. No flu vaccine capability existed, but social distancing and the wearing of masks were regarded as life-saving. Woodrow Wilson left office and was succeeded by Warren Harding. World War I was over, having killed about 20 million including 416 thousand Americans, less than the Spanish flu pandemic. The Confederacy was defeated in the Civil War almost 60 years earlier, but racism persisted via Jim Crow laws and other forms of bigotry. Through the first half of the 1920's, there were about 50 lynchings per year. The Tulsa race massacre of May 31, 1921 killed about 100 blacks and injured 800, in perhaps the worst incident of racial violence in American history. Membership in the Ku Klux Klan surged to over four million. Prohibition had become law in 1919 with the ratification of the 18th constitutional amendment, which unintentionally stimulated illegal production and distribution of alcohol by mobsters. Women gained the right to vote in 1920 under the 19th Amendment.

Your life expectancy was now over sixty years, up nearly fifty percent in just one century. Doctors had learned, among other things, that it was wiser to wash their hands than bleed you. The first antibiotic was discovered in 1909. Women were bearing less than half as many children, but they could hope that all would survive.

The other effects of technology on your life were also transcendent. You could fly in an airplane. You could ride a train from one coast to the other in less than four days. You could buy a Model T Ford for about 300 dollars. You could read at night by incandescent light in your home heated with natural gas. If your husband had gone to San Francisco on business, you could talk to him from your Manhattan apartment via transcontinental telephone. You could listen to Enrico Caruso on your gramophone. You could check your stock prices on the ticker tape. You could see a silent movie, including the famous "Birth of a Nation," accompanied by a live organist. You could shop at your local department store, perhaps Sears Roebuck or JC Penney, and choose from an enormous variety of mass-produced items. The first commercial radio broadcast was on November 2, 1920, from station KDKA in Pittsburgh, but you probably did not have your own radio yet in 1921. On your vacation, you could cruise via Cunard Lines from London to New York in less than five days. In contrast, my 11th grandfather, John Alden, came via the Mayflower in 1620, taking 66 arduous days.

Overall, in the aftermath of the war, pandemic, and stimulated by the Industrial Revolution, the country was poised for one of its most prosperous decades, just before the Great Depression of the 1930's. The twenties are often called "The Golden Age" or "The Roaring Twenties." In comparing the differences in daily lives in 1821 and 1921, you can choose your favorite superlative – unbelievable, historic, epochal, fantastic – you pick. All are true.

Moving forward another century to 2021, the United States population is over 300 million. Human life expectancy has increased to 75 years or so, more than double what it was. Even so, the great COVID-19 pandemic of 2019 is on-going, having killed more Americans than the Spanish Flu a century before and more than the total of American soldiers killed in all wars.

The percentage of the population living on farms has fallen to one percent from the 90 percent or more in earlier times. In other words, we are cumulatively spending less and less time growing our food, even though we are consuming more. You now have a choice of more than 40,000 careers to pursue.

If you are black, brown, or female, you have a chance for great success, even though the playing field is not yet level. Oprah Winfrey, Tiger Woods, Michael Jordan, and LeBron James come to mind as black people in the billion-dollar net worth class. Neil deGrasse Tyson is a principal spokesman for the world of astrophysics. We have elected Barack Obama, a black president, and Kamala Harris, a female multi-racial vice president, and Michelle Obama is often cited as the most admired person in America.

On the other hand, we jail a higher percentage of our people than any other nation on Earth. Black males are incarcerated at more than five times the rate of whites, and there are 2.3 million black males at various stages of detention by the criminal justice system. This is about twice the total number of slaves in American in 1821. A young black male in the United States has a one in three chance of spending time in jail.

If you are female, you can vote and own property, but you might earn substantially less than your male colleague doing the same job. Also, there are plenty of powerful people who wish to dictate your sex and reproductive life. One state legislator has proposed that women who have an abortion should be prosecuted for murder and potentially executed. Alternative sexual orientations and preferences are widely acknowledged and tolerated, even though strongly opposed largely by those who oppose abortion.

Meanwhile though, the industrial and digital revolutions have transformed our lives since 1821 and even 1921. Summarizing it all in a paragraph or two is difficult, but let us try. Television appeared in the late 1940's, and now most of us watch it several hours a day. There are about 3,000 functioning satellites in orbit, including the ones that bring hundreds of channels of television into our homes via 18-inch dish antennas. In addition, we have broadband wi-fi links that let us communicate with the internet as fast as we can type or read. We can hop on a jet plane and reach the other coast in six hours and circumnavigate the world in two days. An astronaut can circle the globe in 90 minutes. Man has stood on the moon and sent space probes beyond the solar system.

To highlight it all, consider the smartphone and its impact. It has been said that "any technology sufficiently advanced is indistinguishable from magic." The smartphone approaches that threshold. Just imagine all that this pocket-sized device does for you for a couple hundred dollars. It makes you infinitely smart, because Google can answer any question you ask. You can satisfy all of your wants and needs by ordering directly from Amazon. Whatever message you care to send to the world, you can do immediately via Facebook or Twitter. You can send a text message, e-mail, or make a telephone call to nearly anyone on Earth. If you have a crusade to initiate, you can write a blog. You can use the GPS system to accurately locate yourself via your map app. You can use the stopwatch to time yourself in a marathon. You can use the camera to take high-resolution photos. You can use a dating service to find a life companion or company for the evening. You can hold a conference call with all of your colleagues while working from home. My smartphone controls my hearing aid and monitors my pacemaker. If you had been able to offer all of these capabilities to Aristotle or Julius Caesar, they might have deified you.

One of the most amazing twenty-first century numbers is that there are over eight million apps for the smartphone. Most provide you with a capability you did not have before.

Forgive me for concluding this chapter with another personal anecdote. My first career step was as a computer hardware designer in 1962. We had developed a line of medium-scale computers for military applications. However, the applications

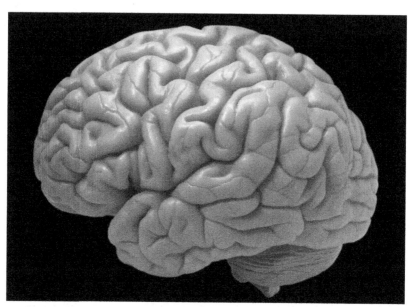

Your brain weighs only three pounds, yet is the most complex known
structure. It has virtually unlimited capacity to retain information.
During your lifetime, you will have over 150 million thoughts.

The smartphone is an example of technology impacting modern lives
by extending our senses, increasing our access to knowledge,
enabling instant communication, and giving us immense capabilities
unavailable to previous generations.

soon outgrew the memory, and so we undertook to develop an extended memory to add another whole megabyte to the storage capacity. Using the latest ferrite core technology, we were able to fit the megabyte into the size of a small refrigerator for only $300,000, or only 30 cents per byte, a bargain at the time

In comparison, last week I needed a flash drive to store my wife's collection of 27 thousand digitally-scanned photographs, and I bought a 128-billion byte drive for $19. At that rate, the same amount of memory in 1962 technology would have cost over $38 billion, a factor of two billion to one. This exemplifies Moore's law, formulated in 1965, forecasting that semiconductor density would increase by a factor of two every year or two, with corresponding decreases in price. Astonishingly, this informal predictor has remained true for the last half century, although it probably cannot be sustained much longer.

You, of course, must reach your own conclusion about life in the twenty-first century, but I suggest that you and I are two of the luckiest creatures ever to inhabit this universe in its entire 13.7 billion years.

Chapter 2
The Scientific Method – How We Know Things

—⋙⋘—

Theories supplied by reason should be verified
by sensory data, aided by instruments,
and corroborated by trustworthy witnesses.
Roger Bacon
(1220 – 1290)

We know that we do not know all of the laws as yet.
Therefore, things must be learned only to be unlearned again,
or more likely to be corrected.
Richard Feynman
(1918 – 1988)

How do we know things? How do we distinguish what is real from what is unreal, and what is true from what is untrue? These are non-trivial questions, and mankind has attempted to answer them as long as we have existed. The principal method devised to address questions of reality is called science. Science, precisely defined, is "the pursuit and application of knowledge of the natural and social world following a systematic methodology based on evidence." The most crucial and difficult part of that definition is the last three words – "based on evidence." For any question, we can ask what is the evidence supporting the answer, and how was the evidence gathered? Is it trustworthy, and can we verify it?

Our first source of knowledge is our parents. We are born knowing essentially nothing, and learn as we go. One of the first lessons is to avoid what might be dangerous. For early man, it was not to pet the sabretooth tiger. More recently, it is not to cross the freeway on our tricycle. Since the dangers to us can be so great, evolution has given us the strong tendency to believe our parents and to follow their direction.

More than avoiding danger, our parents teach us how to behave and what to think. For example, if your parents are confirmed Buddhists, then you are likely to be Buddhist as well and firmly believe its tenets. At some later point, you may choose another system of belief, but you probably will not, and you will regard all other systems of belief as suspect. You will believe what you were taught. You will not depend on evidence to verify your beliefs, but rather on your faith in your parents and your resulting faith in their beliefs. You will then pass these beliefs on to your children to maintain family traditions

Your parents may occasionally teach you things they know not to be true. For example, when you are very young, they may help you enjoy the Christmas season more by teaching that Santa Claus is certain to visit your home on Christmas Eve and leave you presents, providing you behave well and offer him cookies. You may believe this as firmly as anything else your parents taught, at least until the ten-year-old next-door clues you in that there really is no Santa Claus, and also no flying reindeer.

Of course, if you happen to be an eight-year-old math prodigy, you might suspect that something is amiss when you do a little analysis. If there are two billion households on Earth to visit in 24 hours and a cookie weighs two ounces, then Santa descends 23,000 chimneys per second and consumes 250 million pounds of cookies. This seems unlikely. The evidence does not match the story.

The first human families long ago faced similar knowledge problems, but without the benefit of our ensuing experience to draw from. They had to live their lives forward at a time when there was less backward to draw from. They had not yet had time to consider the meaning of life and death, and to understand how the world really works. They knew little more about those things than you did on the day you were born. Survival required most of their attention.

Over two hundred thousand years later, in about 5000 BCE, civilizations emerged. Humans had developed agriculture and slavery, and had imagined an abundance of demons, gremlins, spirits, and gods to help explain existence and its rules. Four millennia later, we reached the Golden Age of Greece, and the life of the philosopher Aristotle (384 BCE – 322 BCE), who is often regarded as the father of science and zoology. He recognized the need to gather data and analyze it as being key to understanding nature. He was the first to classify different animals by their distinct characteristics, and he depended on experiments to study the life forms around him. His influence is profound.

On the other hand, he got many things wrong. He believed that all matter consists of earth, water, air, and fire. He believed that all of space had to be filled with matter or else motion would be impossible. He regarded women as deformed men with fewer teeth, naturally inferior to men, more impulsive, and more deceptive. He believed that men should rule women politically. In these areas, he did not gather enough data to confirm his opinions.

In the two thousand years following Aristotle until about 1500, there were significant contributions to the study of mathematics, primarily in the Middle East and India. Mathematics has the property that much of it can be derived and understood without reference to external reality. A few people suspected that Earth revolves around the Sun, and there were reasonable measurements of the Earth's size. However, insights like these were few and not widely believed, and advancement in what we now call science was sparse.

Progress in science is dependent on three fundamental dimensions. First, there must be people of imagination to conceive what might be real. Homo sapiens provided this from the outset. Imagination led to the creation of civilization in all of its dimensions, for better and worse. Second, there must be recognition that verifiable data has to be gathered and analyzed before reaching any conclusion about what is real and what is not. Aristotle contributed this during the golden age of Greece.

The final fundamental is the development of the technology and tools required to perform measurement and analysis. This advancement accelerated in about 1500, with telescopes, microscopes, better clocks, thermometers, accurate scales, mechanical calculators, the slide rule, the vacuum pump, and others. The invention of the printing press, in 1450, made it possible for knowledge to be documented and communicated directly to others, both in the present and future.

The imagination, the questions, and the tools all came together in the era from 1550 to 1700, which was known as the Scientific Revolution. This period was an element of the Renaissance that marked the transition from the so-called "Dark Ages" to modernity and was characterized by a questioning of old beliefs and the encouragement of curiosity, investigation, and discovery. Key discoveries of the scientific revolution are associated with specific individuals whose contributions have earned their place in history.

Nicolas Copernicus (1473 – 1543) proposed that the planets orbit around the Sun, and that Earth is one of the planets that orbits the Sun annually and rotates daily on its axis.

Johannes Kepler (1571 – 1630) discovered that the planets move in elliptical orbits around the Sun, and he determined other laws of planetary motion.

Galileo Galilei (1564 – 1642) is known as the "father of the scientific method" for numerous contributions, including the use of the telescope to investigate planetary orbits, invention of various compasses, and for championing that Earth revolves around the Sun.

Rene Descartes (1596 – 1650) invented analytic geometry, which unified the fields of geometry and algebra, and introduced skepticism and questioning of results as an essential element of the scientific method. He also significantly impacted philosophy by exploring the relationship of the mind to the body.

Francis Bacon (1561 – 1626) played a key role in extending Aristotelian ideas to encompass the modern definition of the scientific method.

Isaac Newton (1642 – 1727) is often named as the greatest scientist of all time for identifying his laws of motion and gravity, inventing calculus, inventing the reflecting telescope, and studying the nature of light and color.

Although we now speak of the "scientific revolution" and the "scientific method," these terms were not in general use in the seventeenth century, and the field of study of natural phenomena was more typically called "Natural Philosophy." Newton's book on his principles of motion was titled *"Philosophiae Naturalis Principia Mathematica"*, that is *"Mathematical Principles of Natural Philosophy."* This is often considered the most important work in the history of science. The use of the term "science" in the modern sense did not become prevalent until the nineteenth century.

The steps of the scientific method are very straightforward. First, you must have an idea which may be right or wrong. For example, Edwin Hubble suspected, in about 1920, that the universe is expanding. You try to learn what other people believe about the idea and examine their results. Then, you construct a hypothesis about your idea and state it rigorously and concisely. Next, you design and perform an experiment to test your hypothesis. You might decide, as did Hubble, to determine whether there is a red shift in the radiation received from a large number of galaxies, which would indicate they are receding from Earth.

When you have gathered the results of your experiment, you analyze the data objectively and determine whether or not they support your hypothesis in whole or in part. You perhaps need to modify your hypothesis, or gather more data, or throw it all away and seek another idea. If you conclude that the results of your experiments support your hypothesis, then you have it reviewed by other scientists, publish the result, and make all of your experimental data available to any who want to challenge it.

If you are fortunate, you will win acclaim for your work and recognition by having a law of science named after you. That is the ultimate reward for a scientist, and today Hubble's law about the expansion of the universe joins Newton's laws, Einstein's rules of relativity, and Darwin's law of natural selection in this rarified form of immortality. Hubble also has a famous space-borne telescope named after him, which has been one of the most productive scientific tools.

Hubble lived at a time when technology was available to allow his analysis. Otherwise, as with the ancient Greek philosophers, he would have had to guess as well he could. He might have guessed that the universe is eternal and unchanging, or that a god pulls the Sun across the sky in his chariot. If he were persuasive in his story and told it widely, he could perhaps have gotten everyone else to agree with him and even passed laws making it illegal or immoral to believe otherwise.

The burst of scientific knowledge beginning with the Scientific Revolution occurred concurrently with the advent of the modern university. The earliest university still open is the University of Karueein, founded in Morocco in 869. The first university in Europe was the University of Bologna, founded in 1088. Oxford University in England, founded in 1096, is the oldest university in the English-speaking world. Cambridge University came next in England in 1209, formed as a spinoff of Oxford by a group of dissidents.

These two are still among the highest-ranking universities in the world, and together are often called "Oxbridge." Among the notables in science produced by Oxbridge are Francis Bacon, Isaac Newton, Edmund Halley, Henry Cavendish, Thomas Malthus, Charles Babbage, Charles Darwin, James Clerk Maxwell, Ernest Rutherford, Niels Bohr, Edwin Hubble, Georges Lemaitre, J. Robert Oppenheimer, Alan Turing, Francis Crick, Rosalind Franklin, James Watson, Jane Goodall, and Stephen Hawking. This is an unmatched collection. (You can research any of these names you do not recognize. They are all titans.) Together the two institutions have produced 148 Nobel Prize winners.

Today there are over 25 thousand universities in the world, and the fundamental credential that any aspiring scientist must accomplish is the Doctor of Philosophy (Ph.D.) in his or her chosen field. Most leading universities have stated requirements and objectives for their Ph.D. candidates that are the literal embodiments of the scientific method. For example, here are the current requirements for Ph.D. candidates in the School of Clinical Medicine at Cambridge:

At the end of their Ph.D. course, students should: have a thorough knowledge of the literature and a comprehensive understanding of scientific methods and techniques applicable to their own research; be able to demonstrate originality in the application of knowledge, together with a practical understanding of how research and enquiry are used to create and interpret knowledge in their field; have developed the ability to critically evaluate current research and research techniques and methodologies; have self-direction and originality in tackling and solving problems; be able to act autonomously in the planning and implementation of research; and have gained oral presentation and scientific writing skills.

The scientific method is now the world-wide model for extending knowledge. An inherent feature is to regard knowledge as being uncertain and always subject to new results, based on better data and proof. Newton's theory of gravity was modified and extended by Einstein. His theories of relativity have been extended by quantum mechanics. A fundamental goal of science is to extend knowledge in whatever direction the verifiable evidence may lead, however unfamiliar or unpopular.

Modern science can be divided into about twenty major disciplines briefly defined (in alphabetical order) as follows:

Agricultural Science – the study of food and fiber production and processing

Anatomy – the study of the bodily structure of humans and animals

Anthropology – the study of human societies and cultures and their development

Astronomy – the study of celestial objects and the physical universe as a whole

Biochemistry – the study of processes and substances that occur within living organisms

Biology – the study of living organisms and their vital processes

Botany – the study of plants, including their structure, genetics, classification, and ecology

Chemistry – the properties, composition, and structure of elements and compounds

Computer Science – the study of the principles and use of computers

Earth Science – the study of the physical construction of Earth and its atmosphere

Ecology – the relations of organisms to one another and their surroundings

Economics – the study of the production, consumption, and transfer of wealth

Engineering – the design, construction and use of devices, machines, and structures

Geology – the study of Earth's physical structure, its history, and the processes that act on it

Mathematics – the study of number, quantity, and space in their own right and as applied to other disciplines, such as physics and engineering

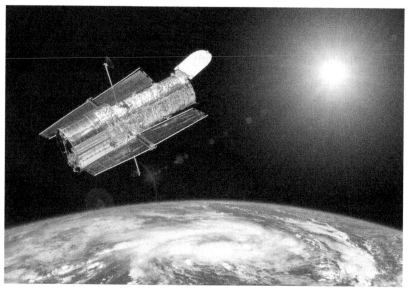

The NASA Hubble Space Telescope was launched in 1990 and is still operational. It has observed galaxies as far as 13.3 billion light years distant.

The Large Hadron Collider (LHC) accelerates protons and ions close to the speed of light. It is the largest and most complex machine ever made with a circumference of 17 miles straddling the French and Swiss borders. In 2012, the LHC confirmed the existence of the Higgs Boson, validating the Standard Model of particle physics.

Microbiology – the study of microorganisms

Paleontology – the study of fossil animals and plants

Pharmacology – the science of drugs, including their origin, composition and use

Physics – the nature and properties of matter and energy

Physiology – the study of the functions of living organisms and their parts

Political Science – the analysis of political activity and behavior, and systems of government

Psychology – the study of the human mind and its functions

Sociology – the development, structure, and functioning of human society

Statistics – the analysis of numerical data

Toxicology – the nature, effects, and detection of poisons

Zoology – the study of the behavior, physiology, classification and distribution of animals

There are many ways to categorize and subcategorize modern scientific disciplines, so the above is not necessarily comprehensive or definitive, but this list spans the vast majority of modern scientific effort and illustrates its scope. Altogether, there are several hundred identifiable specialties of modern science.

Although less than ten percent of humans who have ever lived are still living today, more than 90 percent of all scientists are now alive. Worldwide, there are almost eight million researchers, and over two trillion dollars is spent annually on research and development. The growth of science is universal, and there is much yet to be learned in all fields.

Science has given those of us living today an abundance of verified knowledge that we can use to try to comprehend the life of humanity, both backward and forward, unlike all who preceded us whose only choice was to live forward.

Chapter 3
Numbers Large and Small –
Learning How to Count

Only small minds are impressed by large numbers.
Arthur C. Clarke
(1917 – 2008)

Human beings cannot comprehend very large or very small numbers. It would be useful for us to acknowledge that fact.
Daniel Kahneman
(1934 –)

In the preface, I promised that the ability to read is the first of only two prerequisites to grasping this book. The second is the ability to acknowledge numbers, large and small. If you are among the 40 percent of the population not comfortable with numbers, then this chapter is for you. Even if you are fully numerate, perhaps you will find something of interest. We start with a couple of examples that are both fun and also involve very large numbers.

In 1953, British writer Arthur C. Clarke published a short story, "The Nine Billion Names of God." It is ranked as one of the best science fiction stories of all time. It concerns a group of Tibetan monks who believe that the sole purpose of the universe is for them to make a written list of all nine billion names of God, and that once they have finished, the universe would end. They, however, calculated that it would take over 15 thousand years to accomplish this task. So, the monks rented a computer, had it programmed to complete the job in just a few months, and sure enough, the universe ended. (Perhaps one point of this fictional tale is that we need to be careful how we apply technology!)

If Clarke's monks had instead been given the job of counting all the stars rather than cataloging all the names of God, their job would have been much bigger. A current estimate is that there are about one billion trillion stars in the observable universe. Suppose that we could ordain everyone who has ever existed to be a Tibetan monk with the sole purpose of counting the stars. The total number of humans who have ever lived is about 110 billion. That would mean that every person who has ever lived would be responsible for counting 10 billion stars. If they all lived 100 years, each would have to count 100 million stars per year, or about 200 stars per minute (with no time off for sleeping, eating, gaming, or texting).

Here is another example that illustrates how easily numbers can become incomprehensibly large. Suppose you are a gambler, and you imagine that you can gain an advantage over the house if you have a list of all of the possible orders of a standard deck of 52 cards, from top to bottom, after they are shuffled. The first question before you hire someone to do this is "how many different outcomes are there?" (I am not going to show you how to make the calculation, but it is an easy exercise in combinatorial arithmetic.) For most people's intuition about numbers, the answer is amazing: There are far more different shuffled decks than there are seconds of existence of the universe extending back for all of its 14 billion years! It is almost certain that you could shuffle your deck continuously for your whole life, and never duplicate the sequence unless you cheat. In fact, every human who has ever lived could probably shuffle cards since time began without ever duplicating the resulting order of the deck.

These two examples make a point. If you encounter numbers in any branch of science that are mind-boggling, do not worry about it. They are mind-boggling for everyone. We are simply not equipped to fully comprehend numbers like "a billion trillion." I cannot really internalize a number like that, neither can you, and neither can the smartest person on Earth. I suggest that you just think, "Wow! That's a humungous number," and move on.

Even if you are not comfortable with huge numbers, you probably have some decent skills with numbers and basic lab experiments. For example, consider the bowl of oatmeal you prepared this morning. You measured out the cup of water carefully and decided how hot to make the burner, whether electric or gas. You were careful not to get your fingers too close to the heat. If you were using a gas burner, you might have noticed whether the flame was high enough and adjusted accordingly. Then you measured out the half cup of oatmeal without thinking what it weighed or how many oat flakes there were. When you observed the water starting to boil, you tossed in the oats and perhaps set a timer so you would not undercook or overcook your breakfast. While you were waiting for the cooking to finish, perhaps you added some sugar or raisins to the bowl while thinking about whether or not you were adding too much. You have to watch your waistline, and you do not want too much energy as measured by the calorie count. If the fire was too high, you might have turned it down a little to keep it from boiling over. When you heard the timer go off, you put the cereal in a bowl and added some cold milk from your refrigerator. Finally, you picked up the bowl, grabbed a spoon, tasted and smelled your work and perhaps thought about what you might do differently next time to improve the result.

What happened here from the point of view of the science involved? First, you had to employ your five senses: sight, smell, hearing, taste, and touch. There are no others. Second, you had to ask and answer a number of basic questions: how much, how hot, how long, how many, how far, how often, and perhaps others. But these are the exact same senses and questions used by scientists in generating new knowledge about particle physics or biochemistry. It is just that their tools are more expensive, the problems are harder, and the numbers are much larger or much smaller.

Early humans doubtless needed to count things. The first fossil record of tallies carved in wood or stone dates to about

44,000 years ago. One theory is that the menstrual cycle forced early attention to the lunar cycle of 28 days. The Sumerians are credited with inventing arithmetic, including multiplication and division tables, about six thousand years ago. They had developed a number system for trade and commerce involving sixty symbols. The Romans, of course, used Roman numerals, which are particularly poor for counting large numbers and doing arithmetic. Also, there is no Roman numeral symbol for zero, a hopeless deficiency.

The prevailing system today is Arabic numerals. Using only ten symbols, zero, and one through nine plus a decimal point, allows us to identify any number efficiently. It is widely thought that the correlation between the number of symbols and our number of fingers is not accidental. This system originated in India in about the sixth or seventh century, and was introduced into Europe via Middle Eastern mathematicians, in the 12th century. In terms of their origin, "Hindu numerals," would be a better name and, in fact, the system is sometimes referred to as "Hindu-Arabic." The Chinese had invented a similar system fifteen hundred years earlier, but it did not reach Western civilization prior to the Industrial Revolution.

If we rank the Arabic numerals in order of importance, zero is probably the winner. The notion of "nothing" is usually regarded as one of the most important developments in the history of mathematics. The zero lets us multiply any number one or more by ten simply by writing a zero on the right. Thus one (1) becomes ten (10). I can be more ambitious and go from one (1) to a million (1,000,000) just by adding six zeros (the commas are optional).

This introduces the important notion of the "order of magnitude." That is, if we multiply a number by ten, we are said to have increased it by one order of magnitude. For example, if we go from one to a million, (multiplying by 10 six times) we have increased by six orders of magnitude.

The notion of the decimal point gives us equal facility for numbers less than one. The roots of the modern use of the decimal point again can be traced to Arab mathematicians, but not until the ninth century. Decimal fractions did not reach Europe until the 16th century, or almost coincident with the beginnings of modern science.

The decimal point, together with zero, allows us to decrease any number by an order of magnitude very easily. Thus one (1) can be divided by ten (one-tenth) simply by placing the decimal point to the immediate left of the one (0.1). (We add a zero to the left of the decimal point for clarity.) Similarly, we can go from one (1) to one millionth (0.000001) simply by adding zeroes on the left until we have moved the decimal point six places, thus dividing by 10 six times. By this simple process, we have reduced the number by six orders of magnitude.

This is a great system as long as the numbers are not too big. I can represent my annual salary (say $100,000) or even the value of my house (say $1,000,000) with reasonable comfort. But if we need much bigger numbers (say the number of atoms in a human being), then we would have to carry around many zeroes. Since science has lots of really large numbers, we need a better system. We will encounter some numbers that would require 50 or more zeroes.

Similarly for numbers less than one, we might be comfortable to indicate a penny as one-hundredth of a dollar ($0.01), but other than that we do not want to have to carry around too many zeroes on the left. It is hard to immediately recognize 0.000001 as one-millionth. Science has many cases when we have to represent quantities much, much less than this.

To overcome the problem of too many zeroes, a system called "scientific notation" or "exponential notation" was first devised in Greece around 250 BCE and refined in Europe for Arabic numerals in the seventeenth century. The concept is very simple. Instead of writing a million as a one followed by six zeroes,

(1,000,000), we can write it as 10 followed by a superscript 6, 10^6. The superscript 6 is called an "exponent." If I cannot print superscripts, I can write 10e6 or even 10^6. You will see all of these forms occasionally.

Thus, in the two examples that started this chapter, we could have written "nine billion" as 9×10^9, and trillion billion as 10^{21} (trillion being 10^{12} and a billion being 10^9). Notice that we were able to multiply a trillion by a billion simply by adding the exponents. This is another extremely useful result of scientific notation.

For the record, the number of different shuffles of a 52-card deck is 8.07×10^{67} which is even greater than a trillion billion multiplied by a trillion billion multiplied by a trillion billion. This is a much larger number than we can comprehend.

You might like to master, or at least notice for reference, the scientific notation forms of several commonly encountered powers of 10. These are:

One hundred: $100 = 10^2$ ("ten squared")
One thousand: $1000 = 10^3$ ("ten cubed")
One million: $1,000,000 = 10^6$ ("ten to the sixth")
One billion: $1,000,000,000 = 10^9$ ("ten to the ninth")
One trillion: $1,000,000,000,000 = 10^{12}$ ("ten to the twelfth")
One quadrillion: $1,000,000,000,000,000 = 10^{15}$ ("ten to the fifteenth")

For numbers bigger than these, you might just observe that the number is really humongous, and leave it at that.

Scientific notation is equally powerful for numbers less than one by using negative numbers to indicate successive division by 10. Thus, one tenth is 10^{-1}, one one-hundredth is 10^{-2}, one millionth is 10^{-6}, one billionth is 10^{-9}, and so forth. There are also commonly used prefixes that go with each negative power of 10, and they are worth knowing because they are encountered so often. These are:

One tenth: 10^{-1} (deci) (e.g., decimeter)
One hundredth: 10^{-2} (centi) (e.g., centimeter)
One thousandth: 10^{-3} (milli) (e.g., millimeter)
One millionth: 10^{-6} (micro) (e.g., micrometer or micron)
One billionth: 10^{-9} (nano) (e.g., nanometer)
One trillionth: (10^{-12}) (pico) (e.g., picometer)
(One quadrillionth): (10^{-15}) (femto) (e.g., femtometer)

For numbers smaller than these, just observe that the quantity is tiny beyond imagining. Just for reference, a typical atom is about 100 picometers wide, and we will encounter numbers much smaller than that.

In order to make reading numbers as easy as possible throughout, we will follow current practice and format numbers as follows: Numbers from one to ten are spelled out. Specific numbers from 11 to 99,999 are in Arabic numerals. Numbers of thousands are sometimes written with three zeroes, e.g., 15,000, but the form "15 thousand" may also be used. Numbers involving following zeros are written as hundred, thousand, million, billion, trillion, or quadrillion. Numbers larger than quadrillions use superscript (e.g., 10^{13}), but always with the understanding that they are too big to grasp.

Also note there is no such thing as a largest number or a smallest number. No matter how large, I can always multiply by 10 or any other number. No matter how small, I can always divide by 10 or any other number. Note that the term "infinity" does not refer to a specific number, but to a quantity that has no largest number.

Scientific notation thus gives us the easy capability to cope with numbers, no matter how big or small, over as many orders of magnitude as can be imagined. However, in terms of our understanding of the universe, we are limited by the capabilities of our senses unless we have amplifying tools, such as a microscope. Following are a few comparisons of our senses with what in reality exists in nature.

We need to know about distances; how big things are and how far away they are in space. Our basic human ability in this regard is that we can distinguish things that are as small as 0.1 millimeter (about the diameter of a hair), and we can see nearby stars in the Milky Way galaxy that are about four light years distant. That is an extraordinary skill. However, the smallest distances in the universe that make physical sense are in the range of 10^{-35} meters, and the biggest is the estimated diameter of the universe, which is about 93 billion light years. Thus, our unaided capability to see is more than 40 orders of magnitude less than what might be called 20/20 vision on the eye chart of the universe.

Similarly, we need to be able to measure time as accurately as possible, both in terms of how quickly events happen and over what duration. In that regard, a human can respond no more quickly than about a quarter of a second, and we can personally experience time only over one lifetime, say 80 years. That works out to about 12 orders of magnitude of time. On the other hand, the time for a photon to traverse a hydrogen atom has been measured as about 10^{-19} seconds, and that may be the shortest time ever measured. At the longer end of the time range is the time since the beginning of the universe (about 14 billion years ago), which is about 10^{17} seconds. Our sense of time, as we can directly experience it, is thus many orders of magnitude short of what we can measure given adequate tools.

As a third example, we need to be able to see things as clearly as possible. The universe broadcasts a spectrum of electromagnetic radiation to us in wavelengths as short as 10^{-18} meters and as long as 10^5 meters, a range of 23 orders of magnitude. Our eyes, on the other hand, are tuned to the peak of the light spectrum of our own Sun, and we can see over a range of about 400 nanometers to 700 nanometers, which is a range of less than one order of magnitude. Our brains provide the service of interpreting waves in that limited range as being all of the colors of the rainbow, from red at the longer end to violet at the shorter. We are blind to all but a miniscule fraction of the electromagnetic information that the universe provides. (The parts of the spectrum that we cannot see are called "infra-red" and "ultra-violet").

There are many similar examples with respect to temperature, energy, force, and other technical parameters which all show that our unaided senses are no match for what the universe presents. As Lord Kelvin said, we can fully understand only what we can measure, and the rest is "meagre and unsatisfactory."

This leads us to two essential insights about scientific knowledge. First, before the technological explosion of the last few hundred years, humanity had little chance of correctly figuring out the universe and man's place in it. Although Aristotle, Pythagoras, Archimedes, and others were as smart as anyone, they could only apply their senses and their ability to reason. They could make their best guesses, some good and some not so good. They did not have the tools required to make valid judgments.

Second, modern science is made possible by the development of the tools that enable difficult questions to be asked and answered, supported by experimental results. Observatories, radio telescopes, colliders, particle accelerators, spectrometers, electron microscopes, thermometers, clocks, scales, satellites, computers, and many other instruments have reached capabilities unimaginable even half a century ago. They have revealed both how far we have come and how very far we have yet to go. In many respects, we have only just begun.

Unfortunately, the tools of science can now be prohibitively expensive. For example, the Large Hadron Collider, built in Switzerland largely for the purpose of identifying advanced subatomic particles (including one called the Higgs Boson), cost about $5 billion to build and has an annual operating budget of about $1 billion. An even larger proposed collider, the so-called "Superconducting Super Collider" was a huge device with

A standard deck of 52 playing cards shows how easily very large numbers can occur. The number of different possible shuffled orders of a deck is far greater than the number of seconds of time since the Big Bang occurred 13.7 billion years ago.

Comprehending your universe requires coping with enormous numbers of galaxies, stars, and planets which have existed for billions of years all composed of trillions and trillions of tiny subatomic particles interacting in tiny fractions of a second.

a radius of about 9 miles proposed to be built in Texas. After expenditures of more than $2 billion, the project was cancelled by the United States government due to the projected costs and political issues. At the frontiers of knowledge, the number of dollars required can be as significant as the orders of magnitude of science.

Finally, then, the sum and substance in comprehending yourself in the universe is this: We are all amazing sentient beings with marvelous abilities to see, to feel, and to reason. We can be very proud and grateful for what we can do. The level of comprehension now achieved, however, is largely dependent on the tools invented to extend our senses.

Chapter 4
Roadmap – Milestones of Your Universe and Your World

———ᘓᗄᘓ———

If I have seen further than others, it is by standing on the shoulders of giants.
Isaac Newton
(1643 – 1727)

You are an aperture through which the universe is looking at and exploring itself.
Alan Watts
(1915 – 1978)

In Parts 2 and 3, I describe the sum and substance of events in the order they occurred, rather than when they were discovered. I begin with the creation of the universe, 13.7 billion years ago, rather than when its origin was first understood, merely 100 years ago. From there, I move forward with the progression of events as they occurred, advancing at an astounding average velocity of 50 million years per page!

The progression of the universe is described in Part 2 from its beginning to the first appearance of fully evolved Homo sapiens. If you were to scale all time of the universe's existence into one year, we appear very late in the year. The first Homo sapiens is born at about 10 minutes before midnight on the last day; Julius Caesar and Jesus arrived at five seconds till midnight; and an 80-year human life begins and ends in two-tenths of a second. The description of this vast pre-human time is based on what science currently understands to be the progression of the universe and its component galaxies, stars, planets, and other debris.

Physics is the foundation science that provides the understanding of both the universe and the tiny particles of which everything is made. By 1920, what is now called Classical Physics believed that the universe was reasonably well

41

understood. Atoms were thought to be composed of protons, neutrons, and electrons as the most fundamental atomic particles, and this had explained all 92 naturally occurring elements. The basic principles of electromagnetism were well explained, as were the force of gravity, and the principles of thermodynamics. Some students were advised not to major in physics as there was not much more to be discovered.

Shortly after 1920 and continuing to the present, the apparent certainty of classical physics was shattered in an avalanche of genius. Classical physics has been superseded by Quantum Mechanics. The list of fundamental particles has grown from three to about thirty. The Uncertainty Principle has proven some experimental outcomes are probabilistic and not deterministic. We know electromagnetic waves sometimes behave like continuous waves and at other times like streams of particles called photons. "Quantum entanglement" is the notion that under some circumstances, particles millions of miles apart can act simultaneously. Einstein labeled this "spooky action at a distance." We have proven space, time and gravity are not independent of each other. It has been suggested that a better model than "particles" of the most fundamental atomic reality is "strings", requiring up to 10 dimensions, rather than our three dimensions. We now believe "ordinary matter", or everything making up the universe as we perceive it, comprises only about four percent of the mass and energy in the universe, and everything else is "dark matter" and "dark energy", neither of which is well understood. Finally, it might be that the event creating our universe is not unique, and there may be uncountable other universes. None of this reduces our ability to build satellites and computers or to travel to the planets. It just means, at the deepest level, there is much we do not, and perhaps cannot, fully comprehend.

If you find all of this to be difficult, you are not alone. As the great physicist, Richard Feynman, said, "if you think you understand quantum mechanics, you don't understand quantum mechanics."

The appearance of mankind is of significance to you and me, and Part 3 is devoted to the sum and substance of our world as humans have experienced and created it. The most profound capability of humanity is our ability to alter our environment as well as adapt to it. The average velocity through the book in Part 3 slows down to "only" about 1000 years per page.

Our focus is on several distinct areas. First, we discuss how early humans came to develop civilizations. Next, we examine the various dimensions of civilizations, and what ultimately causes them to end. We note that civilizations usually come into conflict with each other, so warfare is a major topic. Also, virtually every civilization has developed systems of belief, usually called religions, to better understand man's purpose and fate and to govern the behavior of its citizens. We describe some of these systems of belief in their own terms. Finally, the rapid growth of humanity is straining the resources of Earth, and we are in danger of destroying our own habitat. Therefore, Part 3 concludes with a discussion of environmental issues.

Thus, the major areas of focus for our journey through the world are civilizations, warfare, systems of belief, and the environment. You might envision that we are traveling at a high speed down a bumpy road, with lots of twists and turns, ups and downs, and uncertain boundaries towards an unknown destination. Although today, in many ways, is the best time to be alive, the Chinese curse "may you live in interesting times" applies since everything is presently in flux.

How civilizations should best choose to govern themselves is uncertain. Winston Churchill said "democracy is the worst form of government except for all of those other forms that have been tried from time to time." He also allegedly said "the best argument against democracy is a five-minute conversation with the average voter." Following the collapse of the Soviet Union in 1991, the American political scientist, Francis Fukuyama, proposed in his book, "The End of History and the Last Man", that liberal democracy had emerged as the best and final solution to the problem of human government.

This assertion has clearly not been fulfilled, nor is there any prospect of it. The three most powerful nations at this moment are the United States, China, and Russia. The latter two are governed by unrepentant authoritarian autocracies, and the United States is in political turmoil. A recent study, ranking the world's nations by quality of democracy (University of Wurzburg) rates the United States as a "flawed democracy" ranking 36th among democratic nations. About as many people live under autocratic regimes as under democratic ones. Moreover, if there is any trend, it is away from democracy toward autocracy.

Humanity's evolved capabilities in warfare are cataclysmic and unstable. We have the capability to destroy ourselves with either nuclear or biochemical weapons. Warfare in space and cyberspace are also being explored by all of the major potential combatants, and many new weapons are being developed in secret. What is uncertain is the degree to which we will be able to restrain ourselves from general warfare over the long term.

Systems of belief vary widely from region to region in the world. In the aggregate, the five largest systems of belief and their estimated memberships are 1) Christianity – 2.4 billion; 2) Islam – 1.9 billion; 3) Secular – 1.1 billion; 4) Hindu – 1.1 billion; and 5) Buddhism – 0.5 billion. In the aggregate, this sum encompasses about 90 percent of the world's population. However, none is monolithic, and each is divided into many sects. Christians are broadly characterized as Catholic or Protestant, but in detail there are over 10,000 denominations of Christians in the world. Similarly, Islam is divided into two principal Muslim sects, the Sunni and the Shia. There are also similar subgroupings among Hindus and Buddhists. The term "secular" includes all who are not adherents to a specific religion either because of disbelief, uncertainty, or disinterest. Differing systems of belief have the tendency to conflict whenever they come into contact, often resulting either in domination or persecution.

Throughout, we seek to comprehend science, civilization, warfare, systems of belief, and our environment. They interact with each other and determine historical outcomes. For example, Albert Einstein revealed the equivalence of mass to energy in 1903, creating the most famous equation in history, $e = mc^2$. (This equation states very simply that a little bit of mass is equivalent to a whole lot of energy.) Three decades later, he found himself warning President Roosevelt that it was possible to apply this insight to create nuclear weapons. This ultimately proved decisive in ending World War II and determining the balance of power in the world. The invention of nuclear weapons might be the greatest example in history of the law of unintended consequences. Einstein famously said, "If I had known they were going to do this, I would have become a shoemaker."

Notwithstanding all of the uncertainty, at the level of our individual lives, it may be best to follow Voltaire, who asserted ironically that "this is the best of all possible worlds" in spite of all travail. Whatever may happen tomorrow, today is a good day so far. I have a book to write, and you have a book to read. What could be better?

The balance of this chapter is a roadmap of the main milestones through time from the creation of the universe until the present. You might be tempted to skip lightly over this table as you would if you glanced at a paper map before starting a road trip. I encourage you instead to scan these mileposts in order to gain a better perspective on how they fit together. You might even like to come back to it for review when you reach the end of the book.

During his "annus mirabilis" (miracle year) in 1905, Albert Einstein proved the existence of molecules; unified space, time, and gravity in his relativity theory; and showed that mass and energy are related by the equation $e = mc^2$ which led to the discovery and use of atomic energy.

After the development of nuclear weapons, Einstein said "If I had known they were going to do this, I would have become a shoemaker."

MILEPOSTS OF THE UNIVERSE AND THE WORLD

Legend: YA – YEARS AGO; BCE – BEFORE COMMON ERA; CE – COMMON ERA (Note: BCE and CE mean exactly the same thing as BC and AD, just in the format more often used in science today. Dates more recent than 500 CE omit the suffix, CE. Please note that any estimate or date, both in the table and throughout the book, is often an approximation and subject to debate by scholars. I just do not want to repeatedly note "ca." for "circa." That would annoy both of us.)

TIME FRAME   EVENT

From the Big Bang to Homo Sapiens (13.7 Billion Years)
(Reference Chapters 5 – 10)

13.7 billion YA	The universe begins with the cataclysmic explosion of the Big Bang. After initial expansion, the universe begins to cool enabling the formation of subatomic particles followed by atoms.
13.5 billion YA	Milky Way galaxy formed. The home of our solar system containing about 200 billion stars. First identified by Galileo in 1610.
13.3 billion YA	First exploding stars, called supernovae, form the 92 naturally occurring elements, including every atom of you and me. As Carl Sagan said, "The cosmos is within us. We are made of star stuff. We are a way for the universe to know itself."
4.6 billion YA	Our Sun begins to shine. The Sun is classified as a yellow dwarf star formed from the collapse of a cloud of dust and gas called a nebula.
4.5 billion YA	Earth is formed as part of the same process that created the Sun. Initial Earth atmosphere includes almost no oxygen.
3.5 – 4.1 billion YA	First single cell life is born into existence.

3.3 billion YA	Earth's atmosphere becomes rich in oxygen as a result of the appearance of photosynthetic organisms.
2.3 billion YA	First multicellular life appears.
1.4 billion YA	Earth rotates on its axis once in 18.7 hours. Rotation of Earth has slowed ever since to reach the current 23 hours, 56 minutes, and 4 seconds.
580 million YA	Complex multicellular life appears. Sudden diversification of life forms occurs in the Cambrian Explosion which occurred about 541 million years ago.
450 million YA	First vertebrate (species with backbone) exists. Ancestors of sharks and armored jawless fish have been identified through fossil remains.
335 million YA	The Supercontinent Pangaea contains virtually all of the current continents in one land mass. It begins to break apart in tectonic plates to form the current continents about 175 million years ago.
66 million YA	Dinosaurs go extinct, most probably caused by the impact of an asteroid or comet. This paves the way for mammals to become the dominant species. Dinosaurs had existed for at least 230 million years.
6 million YA	First hominid appears. Hominid is defined as all great apes including modern humans, chimpanzees, gorillas, and orangutans and their immediate ancestors. Altogether, there were about 20 hominid species.
3.4 million YA to 3000 BCE	The Stone Age is the long period in which stone is used to make tools and ends with the advent of metalworking (i.e., the Bronze Age).

2.6 million YA to 12,000 YA	The most recent ice age is called the Pleistocene Epoch. During this time, much of Earth's surface is covered with ice.
430 thousand YA	The first known hominid murder is evidenced by a fossil skull which shows two blows to the head.
280 thousand YA	Stone-tipped projectile weapons (spears) permit hunting and mortal combat at close range among hominids.

From Homo Sapiens to the First Civilizations (250,000 Years) (Reference Chapters 10 through 13)

250 thousand YA	Earliest fossil skeletons of modern Homo sapiens come from Africa. Homo sapiens today is the only surviving hominid.
160 thousand YA	Y chromosomal Adam lives in Africa. First common ancestor of all human males.
150 thousand YA	Mitochondrial Eve is the matrimonial most recent common ancestor of all living human beings. She must have existed since mitochondrial DNA is transmitted only from females. She lived in East Africa.
150 thousand YA	Approximate date of the first spoken language, possibly coincident with the emergence of Homo sapiens.
78? thousand YA	Most recent reversal of Earth's magnetic poles. The poles have reversed many times, but science has not converged on how often this occurs and how long the reversals last or on the impact of reversals, up to and including extinction events.
70 thousand YA	Homo sapiens nearly goes extinct as a result of massive volcanic eruption. Estimates of total surviving humans range from a few hundred to a few thousand. This accounts for the very narrow range of DNA among modern humans.

70 thousand YA	Bow and arrow permits hunting and mortal combat at a distance of 100 feet.
50 thousand YA	Humans reach Western Europe. Interbreeding with Neanderthal man occurs. Most large mammal species disappear, probably due to expanding human population. The Neanderthals become extinct by 40 thousand years ago.
40 thousand YA	The oldest known paintings are found on cave walls in Western Europe and Indonesia.
35 thousand YA	First known sculptures. Animal and human figures are carved in bone, ivory or stone. Bracelets and beads also first appear in this period.
30 thousand YA	Before polytheism and monotheism, most early societies believe that many things – animals, plants, geographic features etc. – possess a spiritual essence and influence human affairs. "Animism" is the term generally applied to these beliefs. Many indigenous peoples still practice animism.
14 thousand YA	Humans reach southern tip of South America, the last region except Antarctica to be inhabited by Homo sapiens.
12 thousand YA	First agriculture and domestication of animals. Wild varieties of lentils, peas, and barley are cultivated in Mesopotamia and animals such as goats and wild oxen are herded.
11 thousand YA	Figs are probably the first domesticated plant. Rice is first domesticated in China about 8000 BCE. Domestication of wheat occurs about 7000 years ago in Mesopotamia. Domestication of plants and animals enable the development of civilizations and changes Homo sapiens from hunter-gatherer to farmer and city dweller.

10 thousand YA	The oldest stone oil lamps are discovered in the Lascaux Caves (France) which also include over 600 paintings covering the walls and ceiling.
8 thousand YA	First evidence of irrigation of crops.

From the First Civilizations to Modern Times (6500 Years) (Reference Chapters 13 through 18)

4500 BCE – 1900 BCE	Sumer civilization in Mesopotamia is the first known complex civilization. Sumerians master growing grain and other crops which enable them to form urban settlements.
4500 BCE – 380 CE	Polytheism is the predominant system of belief in early civilizations, notably Egyptian, Athenian, and Roman.
4000 BCE	Dogs are first domesticated in early Mesopotamian civilizations.
3200 BCE	The wheel is first used in the making of pottery in early Mesopotamian cultures. The first use of chariots and carts appears about 300 years later.
3100 BCE – 1200 BCE	The Bronze age is characterized by the use of bronze and other early features of urban civilizations. Bronze is produced by alloying copper with tin and other metals.
3000 BCE – 1100 BCE	The Minoan civilization on Crete, in the Mediterranean Sea, is the earliest advanced European civilization. It features massive buildings, tools, artwork, writing systems, and a network of trade.
2700 BCE	The first war in recorded history occurs in Mesopotamia between Sumer, the earliest civilization, and Elam. It is certain that conflict between adjacent tribes occurred earlier.

51

2334 BCE – 2279 BCE	Sargon of Akkad may have been the first emperor. He conquers all of southern Mesopotamia and establishes the region's first dynasty.
2300 BCE	Hinduism originates in the Indus Valley near modern-day Pakistan. Unlike the other major belief systems, no one person is associated with its founding.
2160 BCE – 1975 BCE	Abraham is the (possibly mythical) patriarch of three monotheistic systems of belief; Judaism, Christianity, and Islam. He is regarded as the founder of monotheism.
1300 BCE	Moses is the legendary central prophet of Judaism, said to have led the Israelites out of Egypt. Biblical scholars debate whether or not he is an historical figure.
563 BCE – 483 BCE	Siddhartha Gautama (the Buddha) is the founder of Buddhism, the world's fourth largest system of belief. He is a philosopher and religious leader who lives in what is now Nepal and India.
480 BCE – 404 BCE	The Golden Age of Athens produces some of the most enduring aspects of Western Civilization including principles of democracy, philosophy, art, and mathematics.
431 BCE – 404 BCE	The Peloponnesian wars between Sparta and Athens result in the complete defeat of Athens and the end of the Athenian golden age.
264 BCE – 146 BCE	The Punic Wars are a series of three wars between Rome and Carthage. Rome ultimately sacks Carthage, slaughters its population, and demolishes it.
4 BCE – ca. 33 CE	Jesus is the central figure of Christianity, the world's largest system of belief. He lives in the Judea province of the Roman Empire which crucifies him at about the age of 36.

1 CE	Human population reaches 150 million.
27 CE – 476 CE	The Roman Empire holds territories around the Mediterranean Sea. Christianity is adopted as the state religion in 380 CE. The fall of the Western Roman Empire is often marked as the start of the Middle Ages.
250 CE – 1539	The Mayan civilization features large-scale construction, large cities, and significant art and intellectual development.
500 – 1500	The Middle Ages (or Dark Ages) generally span the interval between the Fall of the Western Roman Empire and the beginning of the modern era (or Renaissance).
570 to 632	Muhammad is the founder of Islam. He unites Arabia into a single Muslim entity with the Quran forming the basis of Islamic belief.
632 – 661	Rashidun Caliphate expands the Muslim empire. First of the four major caliphates formed after Muhammad's death in 632.
661 – 750	The Umayyad Caliphate is the second of four major caliphates formed after Muhammad's death. After two civil wars, the region of Syria remains the Umayyads' principal power base with Damascus as their capital.
750 – 1517	The Abbasid Caliphs are Arabs descended from one of Muhammad's uncles, and they claim to be the true successors of Muhammad. They invented the world's first paper mill.
850	Gunpowder is invented in China and first used in flaming arrows by the Sung dynasty against the Mongols. The first cannons and grenades are also invented in China.

1095 – 1271	The Crusades include a series of campaigns with the goal of conquering the Holy Land from Islamic rule. This term also describes the effort of Christians to eject Islam from Iberia, which extended to the fall of Granada in 1492. It also applies to various struggles between Catholics and Protestants in Europe.
1206 – 1368	The Mongol empire is the largest contiguous land empire in history and second only to the British empire in total area. At its peak, it stretches from East Asia to Central Europe. It originates with the unification of several nomadic tribes. Its first leader is Genghis Khan.
1264 – 1365	The Yuan Dynasty of China or the Great Yuan is established by Kublai Khan. This dynasty follows the Song dynasty and preceded the Ming dynasty. Buddhism is the state religion.
1300	Artillery permits mortal combat at long distance. First cannons appear in China in 13th century. English cannon first used in the Hundred Years' War in 1346.
1337 – 1453	The Hundred Years' War was the unsuccessful struggle of English royal houses to rule France. This war marks the decline of chivalry and feudalism as well as the rise of nationalistic identities in both countries.
1346 – 1352	The Black Death (the bubonic plague) is the deadliest pandemic in history, killing up to 200 million people in Eurasia and North Africa. The fatality rate is 30 to 60 percent.
1415 – 1999	The Portuguese Empire is one of the longest-lived empires with elements in North and South America, Africa, Asia, and Oceania.

1428 to 1521	The Aztec Empire is an alliance of three city-states in and around the Mexico Valley. They are defeated by the Spanish Conquistadores led by Hernando Cortes.
1438 to 1533	The Incan Empire is the largest empire in pre-Columbian America. It features monumental architecture, an extensive road network, woven textiles, and agricultural innovations.
1440	The printing press is invented by Johannes Gutenberg in Germany. Mass printing enables information to be passed from person to person, place to place, and generation to generation.
1492 to 1976	The Spanish Empire or Catholic Empire is one of the largest empires in history with territories in the Americas, the Philippines, Africa, and Oceania.
1497 to 1945	The British Empire at its peak is the largest empire in history, and for over a century it is the world's foremost power. Its cultural legacy is widespread including to North America, Australia, India and others.
1500	Musket permits hunting and mortal combat at distance of 300 feet.
1500 – 1800	European imperial domination of North and South America. England, Spain, France, Portugal, and Holland bring their weapons, immigrants, religions, and diseases to annihilate 90 percent of native populations.
1576	Built in Denmark, the first observatory enables Tycho Brahe to make comprehensive astronomical observations.
1596	Nicolaus Copernicus (1473 – 1543) of Poland proposes heliocentric (i.e., Sun

at the center) model of the universe. The ancient Greek astronomer, Aristarchus of Samos, had also proposed such a model about 1800 years earlier.

1600   Johannes Kepler (1571 – 1630) develops the laws of planetary motion. He shows that the planetary orbits are ellipses rather than circles.

1615   Galileo Galilei (1564 – 1642) is an Italian astronomer and physicist often called "the father of modern astronomy and physics." His acceptance of a heliocentric (Sun at the center) universe is declared to be heretical by the Catholic Church.

1643   Evangelista Torricelli (1608 – 1647) is an Italian physicist who invented the barometer, which enables the measurement of atmospheric pressure.

1644 – 1912   The Qing dynasty is the last imperial dynasty of China. It was the fourth largest empire in world history.

1676   Ole Remer (1644 – 1710) is a Danish astronomer who was the first to measure the speed of light with reasonable accuracy. His estimate is 131,000 miles per second as compared to the currently accepted value of 186,000 miles per second.

1687   Isaac Newton (1643– 1727) formulates his three laws of motion which define classical mechanics. His laws enable the accurate analysis of the motion of physical objects and systems.

1689   John Locke, "the father of liberalism", publishes his "Second Treatise Concerning Civil Government" and states that all people have rights, including life, liberty, and property.

1698   The first steam powered pump is invented by Thomas Savery for the purpose of clearing flooded mines. Further developments lead to the powering of factories, railroad engines and engines for sailing ships.

1721 – 1917   The Russian empire extends across Eurasia and North America. The Russian senate conferred the title of Tsar on Peter 1, and the Romanov family rules until Nicholas II abdicates upon overthrow by the communists.

1758   The first railway is the Middleton Railway in Leeds, England. All iron rails were introduced in 1796. Steel rails are introduced in the 1860's, and there are soon over 10 thousand miles of railways in England.

1775 – 1783   The American Revolutionary War is waged by the 13 colonies largely in opposition to the costs and taxation of British occupation. George Washington calls the colonists' victory "little short of a standing miracle."

1776   Adam Smith of Scotland, "the father of capitalism", publishes "An Inquiry into the Nature and Causes of the Wealth of Nations."

1778   Antoine Lavoisier (1743 – 1794) is a French chemist who discovered the role of oxygen in combustion. He identifies and names both hydrogen and oxygen.

1803– 1815   The Napoleonic Wars are a series of conflicts between the French Empire and various European coalitions. More than 30 governments are involved and there are over 4 million deaths. These wars have enormous repercussions for the future of Europe and the world.

1816	First working telegraph is built by Francis Ronalds in England. Later enhancements, by Samuel Morse and others, lead to the first widespread use of electrical telecommunication systems.
1835	Human population reaches 1 billion.
1837	Charles Darwin (1809 – 1882) first proposes that all species descended over time from common ancestors by the process of natural selection. This is a foundational concept in modern biology.
1848	Karl Marx, "the father of communism," publishes "The Communist Manifesto."
1860	Dynamite invented by Swedish chemist, Alfred Nobel. Dynamite is the first safely manageable explosive stronger than black gunpowder.
1861 – 1865	The American Civil War is regarded as the first modern war, involving massed armies confronting each other with weapons resulting from the Industrial Revolution. Six hundred thousand soldiers died.
1865	James Clerk Maxwell (1831 – 1879) formulates the theory of electromagnetism and shows that light is a form of electromagnetic radiation.
1867	The periodic table of elements is created by Russian chemist, Felix Mendeleev. It illustrates the chemical properties of the elements in groups according to their atomic weight (number of protons and neutrons.)
1879	The first practical incandescent light bulb is invented by Thomas Edison. Earlier

electric lighting devices had been built by Alessandro Volta in 1800 and Humphry Davy in 1802.

1881	Louis Pasteur (1822 – 1895) is a French chemist noted for discoveries of the principles of vaccination and of pasteurization.
1882	The first hydroelectric power plant is opened in Appleton, Wisconsin. Hydroelectric power currently supplies about 16 percent of the world's electricity.
1903	First powered flight achieved by the Wright Brothers in Kitty Hawk, North Carolina, United States.
1903	First transatlantic wireless message sent by Guglielmo Marconi.
1905	Albert Einstein (1879 – 1956) proves the existence of atoms, defines the theory of relativity, and contributes to the theory of quantum mechanics.
1906	Henry Ford introduces the first mass produced automobiles. By 1927, 15 million Model T Fords are built and the price lowered to $290.
1909	First bomber aircraft designed in England, France, and Germany for World War I. Later bomber aircraft after 1945 permit mortal combat at distance of 9,000 miles and are capable of delivering nuclear weapons.
1913	Neils Bohr and Ernest Rutherford present their model of the atom consisting of a small dense nucleus of protons and neutrons surrounded by orbiting electrons.
1914 – 1918	World War I results in the fall of all continental empires including the Ottoman, Hapsburg, Russian and German. Leads to

the formation of the Soviet Union. Trench warfare and modern weapons yield over 20 million deaths.

1917   The communist revolution overthrows Tsar Nicolas. The practice of religion in Russia is fiercely opposed although never legally outlawed. After the disintegration of the Soviet Union in 1991, freedom of religion was re-established, and almost 50 percent of the Russian population today is Christian.

1918   The Spanish Flu pandemic is the worst since the Black Death, infecting one-third of the world's population, causing 50 million deaths. Note that this is many more than were killed in World War I.

1927   The Big Bang theory postulates that our universe came into existence in one instant of time, 13.72 billion years ago. It is the prevailing model of the observable universe and is confirmed by several independent means.

1927   Television is first demonstrated by Philo Farnsworth. The first television sold commercially in 1929. The estimated total televisions in the world in 2021 is 1.6 billion.

1928   Alexander Fleming (1881 – 1955) is a Scottish microbiologist who discovers the first broadly effective antibiotic which he named penicillin.

1939   John Atanasoff invents the first electronic digital computer at Iowa State College. Key ideas include the use of logic circuits and the binary number system.

1939 – 1945   World War II involves the vast majority of the world's countries directly involving over 100 million combatants. It is the deadliest war in history, with about 80 million deaths, including more civilians than military.

1944   First ballistic missiles deployed by Germany against England in World War II. Modern ICBM's are deployed by seven nations and can deliver nuclear weapons to any point on earth.

1945   Atomic bombs dropped on Hiroshima and Nagasaki, Japan by the United States causing over 200,000 deaths.

1945   Founding of the United Nations (UN). The first global organization of sovereign nations. It currently has 193 member states. Its main stated purposes are to keep peace in the world, to promote friendly relations, and to improve the lives of poor people.

1953   The double helix structure of DNA, credited to James Watson and Francis Crick, but first observed by Rosalind Franklin, is the foundation of modern molecular biology. It has been fundamental in the study of the genetics, modern forensics, and other basic applications.

1954   The first nuclear power plant is the Obninsk Nuclear Power Plant in Russia. Nuclear energy now provides about ten percent of the world's electricity.

1957   Sputnik is the first artificial Earth satellite, launched by the Soviet Union.

1961   The semiconductor integrated circuit is invented by Robert Noyce in the United States. The integrated circuit is at the heart of every electronic device.

1969	Neil Armstrong of the United States becomes the first person to walk on the moon.
1970 – 2020	The Standard Model of particle physics replaces the Bohr-Rutherford Model. Quantum mechanics has yielded significant experimental verification, most recently verifying the Higgs Boson particle at the Large Hadron Collider in Switzerland.
1983	The communications method which allows multiple computers to interact was invented by a US Government agency. This formed the basis of the internet.
1992	The first commercially available device that could be called a smartphone is developed at IBM.
1996	Dolly, a sheep, is the first mammal to be cloned from an adult cell at the Roslin Institute in Scotland.
1997	The first social media site, Six Degrees, enables users to load a profile and make friends.
2019 – ????	The COVID-19 pandemic is the largest since the Spanish Flu of 1918. As of February, 2022 there have been over 400 million infected and over six million deaths.
2021	First flight of helicopter on Mars. United States, China, and Russia are all active in solar system exploration by unmanned vehicles. Number of operational satellites in orbit exceeds 3000.
2021	James Webb Space Telescope launched to a point 9 million miles from Earth.
2021	Human population estimated as 7.9 billion less than 200 years after reaching one billion. Global life expectancy is 72.6 years, up from about 30 in prehistoric times.

Part 2
Comprehending Your Universe

Chapter 5
The Universe – BANG! You Are Alive

The universe is not only queerer than we suppose,
but queerer than we can suppose.
J. B. S. Haldane
(1892 – 1964)

Great fleas have little fleas upon their backs to bite 'em
And little fleas have lesser fleas,
And so, ad infinitum.
And the great fleas themselves,
In turn, have greater fleas to go on;
While these again have greater still, and so on.
Augustus De Morgan
(1806 – 1871)

Prior to the scientific revolution, all cultures conceived stories of how the universe and Earth came to exist. These stories are called "creation myths," and there are hundreds of them. The oldest written creation myth, originating in Babylon four thousand years ago, is called the "Enuma Elish." The story involves a great battle between gods Marduk and Tiamat that results in the creation of Earth and mankind. Contemporary religions generally regard their creation stories, such as Adam and Eve in the garden of Eden, either as literally true or as symbolic narratives containing significant truths.

This chapter tells the creation story of the universe as currently understood by science. Prior to its being, space and time did not exist for the universe. If anything else existed earlier, it is unknown. Then, in the briefest instant, what is called a "singularity" came to be, and the singularity was incredibly dense and hot, and it included all of the mass and energy observed in our universe today. We estimate to a high degree of accuracy that this event occurred 13.72 billion years ago. Initially, the universe was so dense and hot that ordinary matter could not exist; instead, there were only energy and subatomic particles. Then an incredible period of expansion and cooling began, ultimately leading to the formation of atoms, elements, stars, and galaxies, continuing far into the future.

The creation of the universe can now be analyzed, understood, and, based on new evidence, modified. Before about 1920, the tools and science to discover its origin had not yet been invented.

This creation explanation came to be called "the Big Bang" which is inelegant, but at least descriptive. The name was first coined in the 1920's by astronomer, Fred Hoyle, who did not believe it, but instead favored the theory of a static and unchanging universe. Why should we believe the Big Bang story? After all, it seems unimaginable that all of the mass and energy now existing in the cosmos could have been jammed into one infinitesimal point. You might wish that there was a better explanation.

Recall what was believed in Athens during its golden era about 2500 years ago, at the time of Aristotle, Socrates and Plato. The Greek pantheon of gods included the sun god Helios who was represented as a young man with a golden crown who drove the chariot of the Sun across the sky each day and returned through the world-ocean at night. Helios was worshipped, temples were built and sacrifices were offered. Many Athenians believed in him without doubt. He later was carried forward to the Roman pantheon as the sun god, Sol, and there are sun gods in numerous other early systems of belief.

Why should we believe the incredible story of the Big Bang in favor of the comparably incredible stories of Helios and Sol? The answer is a fundamental one in comparing the scientific method with all other systems of belief: The Big Bang story is accompanied by a substantial body of experimental evidence which has been tested and retested by thousands of scientists over the past 100 years.

In the case of the Big Bang, there are three fundamentally independent verifications that the universe began in one cosmic event. The first is a consequence of the fact that all galaxies in the universe have been shown to be moving away from us at an increasing rate. Space itself is expanding and all galaxies are becoming more and more distant from each other. Thus, if you look backward in time rather than forward, everything in the past would have been closer together than now, and in fact, at some measurable time in the past, must have all been very nearly in the same place. This is not necessarily a decisive argument, but there are two other results which validate and confirm the Big Bang story.

We know that if the universe was denser and hotter than it is now, the heat would have been sufficient to generate electromagnetic radiation. Given our understanding of the process, the temperature in the universe shortly after the Big Bang was an enormous 10^{32} degrees Centigrade. We should thus expect to be able to detect the radiation that was transmitted uniformly into the entire universe just as we detect radiation from all galaxies we can see. The magnitude of this radiation as seen at present day Earth was forecast in about 1940 and confirmed, in 1964, by measuring the radiation exactly as predicted, thereby affirming the Big Bang.

The third confirmation of the Big Bang results from modern quantum mechanics. If the universe was initially very hot, very dense, and expanding, then it is possible to show that initially atoms cannot exist, and also to predict the ultimate timing, composition and distribution of matter into elements. For example, the visible mass of the universe today consists of about 73 percent hydrogen and 25 percent helium, the two lightest gasses, exactly as predicted. Everything else, all 92 elements, only constitute about two percent and was formed much later in the tremendous explosions of dying stars.

Then, given the understanding of the beginning, we are able to grasp what came after. Physicists can describe the first instants of the universe in excruciating detail down to the smallest fraction of a second, albeit with some uncertainty and alternative viewpoints. There is consensus that the universe began at a point and initially was nearly infinitely dense and hot. It then began to inflate at an amazingly high rate (called "cosmic inflation"), and began to cool. As it did, all of the mass and energy evolved in various stages to the subatomic particles as we know them today. Only twenty minutes or so after the Big Bang, the reduced temperature and pressure of the universe yielded atoms of light elements in the proportions of about 75 percent hydrogen, about 25 percent helium, and tiny fractions of a percent of lithium and beryllium. Essentially, everything was hydrogen or helium and the temperature was too low to enable formation of any of the heavier elements.

The pace of things then slowed down a lot. The first molecule (that is, a uniting of hydrogen and helium atoms) did not occur until about 100,000 years after the Big Bang. In fact, there was a long period of several hundred million years, called the Dark Ages, when there was no visible light in the universe. Finally, hydrogen and helium began to form masses, called stars, within a few hundred million years after the Big Bang.

Groups of stars aggregated to form the earliest galaxies, between 400 million and 700 million years after the universe began. A galaxy is an enormous collection of gas, dust, and billions of stars and their solar systems, all held together by gravity. Galaxies are generally classified by their shape. The three general types are elliptical, spiral, and irregular. (Our Milky Way galaxy is a spiral.) As the galaxies emerged, the universe became filled with their light. The Dark Ages ended after one billion years, and the universe took on its present appearance with billions and billions of galaxies receding from each other, more and more rapidly. This is our universe as we find it now, 13.7 billion years later.

Space-based telescopes have advanced so that we can see radiation from the earliest galaxies formed over 13 billion years ago. As a result, the estimated number of galaxies continues to increase. Less than a century ago, we thought the Milky Way was the only galaxy and the total extent of the universe. At the turn of this century, only two decades ago, that number had increased to 125 billion galaxies. The most recent estimate is that there are two trillion galaxies – over two hundred for every person on Earth! Nearly unimaginably, the average number of stars per galaxy is perhaps 100 million, with a known range of one thousand to 100 trillion.

It may not have occurred to you that the study of light from the universe is the study of very ancient history. In other words, if we are looking at an image through our telescope that came from a star 10 billion years old, then we are looking at photons that started on their journey to our eyes 10 billion years ago. At the speed of light (186,000 miles per second), a photon which started toward us 10 billion years ago has traveled almost 6×10^{22} miles.

If you were able to collect an average sample of the universe as it exists today, you would quickly discover two things that might surprise you. The universe, on the average, is very empty and very cold. The density is less than 0.25 atom per cubic meter and the average temperature is about -270 degrees centigrade. This is less than three degrees above the lowest temperature possible, the temperature at which virtually all atomic motion ceases. Moreover, since the universe is expanding, the average density and temperature will continue to decrease.

Astronomy would be a very uninteresting science if it only collected data on empty space. Instead, it studies galaxies, stars and everything that produces light, gravity, and motion. The process of galaxy and star formation has been continuous to this day. Smaller galaxies are routinely swallowed by larger ones. Across the universe 4800 new stars are born every minute, although the rate of new star formation is decreasing as the universe continues to expand and become cooler. However, you can contemplate that, somewhere out there, a new star similar to our Sun is being born today, and five billion years from now it may host a planet similar to ours.

Stars have a finite lifespan, and they all go through the same four stages. In the first stage, as a result of gravitational attraction, a large mass of hydrogen and helium coalesces from the vast clouds of matter, called nebulae, which resulted from the Big Bang and ensuing stellar explosions. Gravity allows the star to accumulate matter, becoming larger, denser and increasing in temperature, but not yet hot enough or dense enough to enable the fusion of hydrogen atoms to begin. This first stage lasts for perhaps 100 million years. In the second phase, the developing star continues to attract nearby mass due to gravity. It begins to rotate, and the momentum of this rotation casts off the matter which will become most of that star's solar system – planets, asteroids, comets, and the like. This second stage also is typically about 100 million years in duration.

In the third stage, the mass, density, and temperature enable nuclear fusion of hydrogen into helium to begin. The star is born and enters what is called its "main sequence" after a gestation of only 200 million years. A star is then effectively a continuously exploding hydrogen bomb producing immense amounts of energy and radiation through fusion of hydrogen into helium. The length of time of the main sequence depends almost entirely on the initial mass of the star.

Our very own star, the Sun, has been in its main sequence for about five billion years and it will continue for another five billion or so. In astronomer's terms, the Sun is classified as a "main sequence yellow dwarf" star. This happens to be one of the most frequently occurring type of star in the universe. (We will discuss the details of our Sun's lifespan in Chapter 7.) For convenience in comparing other stars, the Sun is designated to have one solar mass, and other stars are categorized in terms of their number of solar masses. (For the record, the mass of the

Sun is about 2×10^{30} kilograms, another one of those humungous numbers that cannot be grasped.)

The heaviest star we know has over three hundred times the mass of our Sun, and the smallest has only 0.07 solar masses. Heavy stars have the shortest main sequence lifetime and lightest stars the longest. Heavy stars reach much higher internal temperatures because of the greater force of gravity resulting from their large mass, and therefore burn faster. The largest stars shine for only about a million years while the smallest will shine for up to ten trillion years, or seven hundred times the lifetime of our universe to date.

In addition to coming in a large range of masses, stars also come in a wide variety of size, color, temperature and luminosity. (Luminosity is a measure of the amount of light that a star produces.) One of the largest stars, called red giants, would fill our solar system all the way to the planet Jupiter. That is a radius of about a billion times the radius of the Sun. The smallest stars, called white dwarfs, have a radius of about one hundredth that of our Sun. Colors range all the way across our visible spectrum, from blue to red. Our Sun is yellow. The color and luminosity of a star primarily depends on its surface temperature which ranges from about 2500 degrees centigrade for a red dwarf all the way up to 25,000 degrees centigrade for a white dwarf. The surface temperature of our Sun is about 5,800 degrees centigrade.

Inevitably, in the fourth stage, all stars come to the end of their main sequence and cease to shine when their supply of hydrogen to fuse into helium comes to an end. They burn out, but as they do so, they go through a predictable sequence of dying stages. The final sequence varies primarily depending on their total mass.

An especially interesting and dramatic final stage for some stars is to bow out in gigantic explosions called "supernovae." These occur in several ways for stars of varying sizes. The simplest to understand is for a very massive star. When its main sequence comes to an end, the internal temperature falls because the supply of hydrogen to fuse to helium has been depleted. When that happens, the inward force of gravity due to the mass of the star overcomes the outward force due to fusion, and the star collapses inward in a very short period of time. This generates enough heat that the entire star explodes and disintegrates, blasting most of its mass into space forming nebulae. The temperature can reach over one billion degrees centigrade. The collapse of the core can take as little as a quarter of a second, and the energy emitted can exceed the entire energy of our Sun over its entire life. The supernova will be incredibly bright for only a few days and will fade away in just a few months or years.

Supernovae which occur within our Milky Way galaxy are visible with the naked eye and are often the brightest objects in the sky, except for the Sun. They have been recorded as notable historic events, with occurrences in 185 CE, 1006, 1054, and 1604. They do not occur often in our galaxy, and none has been observed since 1604. With modern telescopes, however, supernovae are detectable in more remote galaxies, and are now noted several hundred times per year. Our Sun is not a candidate to become a supernova because it is neither massive enough nor hot enough.

An important thing to know about supernovae is that without them, human life would not have occurred. We noted earlier that hydrogen and helium were almost the only elements created in the Big Bang. The high temperatures associated with supernovae enable the fusion of lighter elements into the iron, silicon, gold, silver, and other heavier elements without which neither the planets nor life is possible. All of this material blasted into space by supernovae joins the hydrogen and helium from the Big Bang in forming the nebulae which are the source of everything that comprises all of the solar systems in the universe.

A possible final outcome for the supernova is for any remaining mass which is not blasted away to be so dense that not even light can escape it because of the enormous pull of gravity, thus forming what is called a "black hole." A teaspoon of typical black hole material would weigh about 10 million tons. Once a black hole is formed, its enormous force of gravitational attraction can enable it to swallow up anything that comes close. As a result, supermassive black holes have formed at the center of some galaxies, including the Milky Way. Complete understanding of black holes is still forming, with some evidence that black holes are at the center of most galaxies, and there may be a strong relationship between the formation of the black hole and the galaxy itself.

Given all of our knowledge of the galaxies and their components, we can estimate total mass-energy and predict their trajectories as governed by the force of gravity and the laws of motion. When this is done, we discover that our observations of galactic motion do not come close to our predictions. The galaxies are rotating at a faster rate than predicted. The conclusion is that there must be some force not accounted for by the mass of the matter that we can see. We call all of the observable mass "ordinary matter", and we call the matter that we cannot see using current technology "dark matter." (The terms "transparent matter" and "unexplained matter" might be as descriptive, but "dark matter" is the name that has come to common usage.)

It is surprising and challenging that the total of dark matter required to accurately calculate galactic trajectories is about six times as great as all of the ordinary matter. There is a lot more stuff out there we cannot see than stuff we can see. A favorite theory at the moment is that there is some fundamental particle which causes the unexplained force and which is invisible to all of our senses and which we have not yet learned to detect. Nobody can explain it very well yet, but many are trying.

Our understanding of the universe remains incomplete. For most of the last century, physicists have debated its ultimate fate among three alternatives. First is that although the universe has expanded since the Big Bang, the force of gravity will cause the expansion to slow and ultimately reverse. In that event, the universe might again converge on one point (commonly called "the big crunch") and that this process might endlessly repeat itself. The second possibility is that the expansion of the universe will eventually come to a stop at some equilibrium and remain in a static state forever. The third possibility is that the universe will continue to expand without limit.

It now appears that this question has been settled based on measurements by modern telescopes. Apparently, the universe will continue to expand without limit and moreover it is expanding at an increasing rate. As for dark matter, it is again amazing and perhaps a bit discouraging, that the total energy required to cause the expansion we observe is far greater than the energy we can identify. This energy has been named "dark energy" which equates to "unexplained energy." The favorite theory of the moment suggests that there is a small component of energy, constant per unit of volume of space, and as the universe expands, total energy continually increases.

The current estimate is that of all of the mass-energy in the universe, 68 percent is dark energy, 27 percent is dark matter, and all of the ordinary matter we have studied for so long accounts for only five percent of what is really out there.

As of now, no one can explain it very well, and we do not know if and how the expansion of the universe ends. Recall from Chapter 3, that as far as subatomic particles go, we also do not know if we have found them all or even whether atoms are finally best understood as ten dimensional strings rather than as particles.

So, at the end of science's creation story, you and I are as fleas upon Earth; and Earth is as a flea upon the solar system; and the solar system is a flea upon the Milky Way galaxy; and the Milky Way galaxy is as a flea upon the cosmos; and above that, nobody knows for sure.

Yet our living cells are as fleas upon us; and our DNA is as fleas upon our cells; and the molecules are as fleas upon the DNA; and the atoms are as fleas upon the molecules; and the subatomic particles are as fleas upon the atoms; and below that nobody knows for sure.

At both ends, reality may be beyond our ability to know. The wonder of it is that humans are the only fleas that comprehend the others. Their purpose may be as great as ours, but we are smarter. We have come from believing that the Sun was towed across the sky by a god in a chariot, to being able to describe the beginning of our universe from the smallest fraction of a second. To be sure, there are basic things that we do not yet grasp, both at the subatomic and intergalactic levels, but we have come a long way, and there are still many unknowns.

Chapter 6
The Milky Way – A Galaxy of Stars

—◦◦◦—

In less than a hundred years, we have found a new way to think of ourselves. From sitting at the center of the universe, we now find ourselves orbiting an average-sized sun which is just one of millions of stars in our own Milky Way galaxy.
Stephen Hawking
(1942 – 2018)

We all travel the Milky Way together, trees and men.
John Muir
(1838 – 1914)

Before we begin our conversation about the Milky Way Galaxy, please open the YouTube application on your TV, your smartphone, or your computer. Then search on "Hubble Telescope Images", and spend half an hour or so looking at the images. You might also spend another half hour on a photo show of the Milky Way. Take time to contemplate their beauty and their enormity, and stop to realize that they represent what is really happening while you are having dinner. Allow your mind to process the images in whatever way it cares to. At the end of this brief experience, you will know more about and have seen more of the galaxy than anyone prior to the twentieth century.

When we left the universe at the end of Chapter 5. The Big Bang had occurred and blasted an enormous cloud almost entirely of hydrogen and helium into rapidly expanding space. Due to small perturbations in the nearly constant density, perhaps in part because of what we now call dark matter, the hydrogen and helium began to form large clumps. As each clump became larger, the force of gravity at its center caused the internal temperature and pressure to rise. Finally, hydrogen atoms began to fuse into helium atoms accompanied by a large release of heat and energy, and the clump lit up to become a newborn star. The total life story of the star was then largely pre-determined by its initial mass and temperature. Then, because of gravity, large numbers of stars clustered together to form galaxies, yielding the two trillion galaxies which we can now see, thanks to tools like the Hubble Telescope.

One of those galaxies, first named by the ancient Greeks, is our home – the Milky Way. All of the stars you can see with the naked eye are in the Milky Way. Other early civilizations also noted the bright area in the sky and attached their own mythology to it. Hindus called it "the Ganges of the Sky" while in China it was named "the Silver River" and in Japan the "River of Heaven."

The basic shape of our galaxy is a spiral with four main arms. Our Sun is located about 26 thousand light years from the center of the galaxy, in one of the spiral arms. There is nothing particularly remarkable about the Sun, except that it enables our life, making it very remarkable to us. The details of the galaxy are presently the subject of intensive research enabled by space-based telescopes, and knowledge in this area has grown rapidly in the last two decades. As with the physics of particles and the universe, there is much yet to be learned.

The Milky Way is one of the oldest galaxies in the universe, with an estimated age of 13.5 billion years, formed "only" 200 million years after the Big Bang. One verification of this is that over 40 stars in the Milky Way have been identified by spectrum analysis as being comprised almost entirely of hydrogen and helium. This means they were formed very early, before heavier elements existed. By contrast, the newest galaxies in the universe have formed within the last 500 million years, and contain a larger proportion of heavier elements.

A problem in determining the configuration of our galaxy is that we live in the middle, with no way to take a picture of it all. The nearest point in space from Earth which would allow a photo of the galaxy to be taken is several hundred light years away. Another basic problem in investigating our own galaxy is that it is impossible to look through the center to the other side, since the center consists of a very high density of light and dust.

By contrast, determining the spherical shape of the Earth is much easier than determining the shape of the Milky Way for several reasons. First, we have observed since ancient times that the Sun, the moon, and the planets are all round, so it was a reasonable guess that Earth is round as well. Then, Magellan's voyage circumnavigated the globe in 1522. Finally, we sent astronauts to the moon, and they sent back the famous photo of our beautiful blue, and spherical, planet. The physics is also easy to understand, since as gravity is attracting a large mass to a common center point, then in the end every point on the surface will be at least approximately equidistant from the center, i.e., a sphere.

From Earth, on a clear night, we can look into the sky and observe one region which appears to be brighter than the rest. This is one of the four spiral arms of the galaxy. (Currently, because of light pollution by cities, fully one-third of Earth's population cannot even see this region of the Milky Way.) The ancient Greeks interpreted the display as breast milk spilled by the goddess Hera, thus giving the galaxy its name. With modern telescopes we can see the shapes of innumerable galaxies in space coming in a variety of colors and shapes - ellipses, spirals, and shapeless blobs. We have observed that over 70 percent of galaxies are spirals with a central bulge surrounded by a flat rotating disc of stars, making this the likeliest form for the Milky Way, which has since proven to be correct.

There are two main areas of investigation that allow us to determine conclusively the basic shape of our galaxy is a spiral, although there is still substantial disagreement about the details. First, by observing the heavens from points all over our planet, we can see the distribution of stars is not uniform as it would be if the galaxy was more uniform than a spiral. Second, we can determine that the velocities of stars and gas in the galaxy show a non-uniform rotational motion typical of spiral galaxies. Modern astronomers have gathered montages of thousands of images taken from and around Earth, enabling detailed analysis of the size, shape, composition and evolution of the Milky Way.

Here is a summary of the current understanding of the Milky Way. It is a spiral disc with four arms. The galaxy has a central bar with a supermassive black hole at its center and a high density of stars and dust. The estimated total number of stars in the galaxy is 100 billion to 400 billion and at least that number of planets. (The number of stars is so uncertain because of the difficulty in identifying brown dwarf stars.) The diameter of the stellar disc is about 120 thousand light years and the thickness of the disc is about two thousand light years, or only about one percent of the diameter. The galaxy's range of gravitational attraction includes a halo with diameter of about 1.9 million light years, or about ten times the stellar diameter. Analysis of the rotational period indicates that up to 90 percent of the mass of the Milky Way is dark matter, as compared to only 65 percent for the universe as a whole. The galaxy has a rotation period of about 240 million years, which means there have been only about 20 revolutions since Earth was formed and only about one-thousandth of a rotation since humans evolved.

Galaxies and stars have fundamentally different life cycles. A star is formed when sufficient mass accumulates to enable fusion. It burns until its fuel is exhausted, and its life cycle ends. Galaxies, on the other hand, are accumulations of many stars and nebular dust. New stars are continuously being formed amid the nebulae, and new nebulae are created with the supernova explosions accompanying the end of some stars' life cycles. At present, the Milky Way is estimated to be creating new stars at the rate of about three solar masses per year. However, the total rate of new star formation is decreasing due to the gradual exhaustion of available gas. Simulations suggest new star formation in the Milky Way will end about five billion years from now. Other, younger, galaxies will continue to form new stars for billions of years more.

Galaxies tend to aggregate into clusters called "local groups." The Milky Way is part of a local group including more than 30 galaxies spread over a diameter of about 10 million light years. The Milky Way, as presently configured, is the result of absorbing at least ten smaller galaxies over the course of time. The three largest galaxies in the Milky Way group today are the Andromeda Galaxy (the largest), the Milky Way (second largest) and the Triangulum Galaxy. Galaxies attract each other as gravity dictates, and galaxies occasionally collide to form a newer, larger galaxy. This is often the cause of spiral galaxies evolving into elliptical galaxies after intergalactic collisions occur. Most notably, the Milky Way and Andromeda are on a collision course, which will probably bring them together in approximately five billion years and will distort the configuration of both, until a new galaxy is established in about 10 billion years. Astronomers are not yet sure what the configuration of that galaxy will be.

Large galaxies, including both the Milky Way and Andromeda, have supermassive black holes at their center. Sagittarius A, at the center of the Milky Way, was initially thought to be a very bright star. However, recent analyses of the orbits of nearby stars have proven that Sagittarius A is actually a black hole about four million times more massive than our Sun, accompanied by a very bright nearby star which has not yet been absorbed into the black hole. The mass of Sagittarius A will continue to increase as nearby stars and other matter are sucked into the black hole by its immense gravity.

An obvious concern about black holes is that they might absorb all the mass of the universe. You might be relieved to know that physicists now believe there is a maximum size for black holes in the range of 50 billion solar masses. The basic reason is that black holes finally absorb all of the nearby gas, and then there is no further mass within the black hole's range of gravitational attraction.

Another star of interest in the Milky Way is the red giant, Betelgeuse, (pronounced "beetle juice") which is a candidate to be the next supernova in the Milky Way. As noted earlier, the most recent supernova occurred in 1604. Betelgeuse is one of the brightest stars in the night sky and is also one of the largest with a radius of 360 million miles, almost 900 times that of our Sun. (The largest known star in the universe has a radius over 2000 times that of the Sun.) If Betelgeuse were in the Sun's place, it would extend roughly to the orbit of Jupiter. It is a good example of a star with a brief life span, less than 10 million years, because of its large mass and resulting high internal temperature. It is almost certain to end its life cycle in a gigantic supernova explosion. Since it is "only" 650 light years away from Earth, it will be one of the brightest events ever witnessed by mankind. However, it will be extremely bright for a very short time – perhaps a month or two. The explosion itself will take only a couple of minutes to occur. Of course, we will not see it until 650 years after it actually happens. It is expected to be visible in the daytime and as bright or brighter in the night sky than a full moon. Astronomers maintain a very close eye on Betelgeuse, hoping that the supernova might occur in our lifetime. Sadly, the best guess is that this supernova will probably not happen for another hundred thousand years or so.

"Does intelligent life exist elsewhere?" is another fascinating question that arises as a result of realizing how many galaxies and stars there are. The number of planets which orbit stars in the Milky Way is at least 100 billion, and the number of planets orbiting stars in the universe is 10^{25} (another one of those unimaginably large numbers, but still far less than the number of different shuffles of our 52-card deck).

An on-going project, the Search for Extraterrestrial Intelligence (SETI), has been underway for over 50 years, with

Our Milky Way galaxy is a spiral disk that measures about 120 thousand light years across and rotates once every 240 million years. It includes over two hundred billion stars, but about 90 percent of its mass is unseen dark matter.

All of the stars we see with the naked eye from Earth are in the Milky Way. Light from the center of the galaxy takes 26 thousand years to reach us.

no conclusive evidence yet. Modern observatories have enabled the identification of about seven thousand specific exoplanets, defined as "planets outside our solar system." Their principal technique is to measure small variations in the total light output of a star resulting from a planet passing in front of it. A recent estimate is that there are over 300 million potentially habitable exoplanets in the Milky Way. On one hand, many cosmologists believe that because there are so many potentially habitable planets in the universe, intelligent life must have evolved many times. On the other hand, many unlikely conditions must exist to enable life, and as physicist Enrico Fermi said, "if the universe is teeming with aliens, where is everybody?" This is called "the Fermi paradox." So far, nobody knows the answer, because we have not found anyone and no one has found us. This remains a great subject for science-fiction writers and for conspiracy theorists, as well as for scientists.

Chapter 7
The Sun - The Light of Your Life

————∞————

One can understand why many cultures through the millennia have worshipped the Sun.
Stephen Poplin
(1954 –)

The Sun, with all those planets revolving around it and dependent on it, can still ripen a bunch of grapes as if it had nothing else in the universe to do.
Galileo Galilei
(1564 – 1642)

In our journey together in time and space, we now take the next huge step toward home. The background radiation from the Big Bang, currently arriving at your satellite dish, is 13.7 billion years old. Starlight from the center of the Milky Way, near Sagittarius A, started toward Earth 26,000 years ago, about the time of the first cave paintings in Europe. The light from Betelgeuse began its trip to your eyes shortly before Shakespeare wrote Romeo and Juliet. However, the sunlight illuminating your world today took only about eight minutes and twenty seconds, traversing the 93 million miles between Sun and Earth at the speed of light, 186,000 miles per second. The distance from the Sun to Earth is defined as one astronomical unit (AU), which is convenient in grasping distances between things in the solar system. If you were to travel to the Sun at the same speed that Neil Armstrong went to the moon, the round trip would take ten months. Upon arrival, you would find that the surface temperature is almost 6000 degrees Centigrade.

Life on Earth is utterly dependent upon the good behavior of the Sun. The planet we live on was formed during the process that created our Sun, and both occurred at about the same time, over four billion years ago. All plant life on Earth uses energy radiated from the Sun to create organic matter through photosynthesis, the process that plants use to make energy-rich organic compounds and oxygen. Herbivorous animals derive all of their energy from the plants they consume, while carnivores derive their energy from consuming other animals. Humans, as omnivores, have the choice of consuming plants or animals, although as population increases, it is important to note that consuming plants is about 90 percent more efficient than consuming animal flesh. Hydroelectric power depends on the Sun to drive the global cycle of evaporation and precipitation that fills the rivers which are dammed for hydropower generation. Plant photosynthesis also generates our oxygen-rich atmosphere, so the air we breathe, the food we eat, and the energy to power our appliances, all derive directly from sunlight. Fossil fuels (oil and coal) are simply ancient organic matter, once dependent on the Sun for life. Many of us are powering our homes with solar panels and are also using them to recharge the batteries of our electric cars.

By the time of the earliest civilizations, Homo sapiens needed to understand and attempt to influence the behavior of the Sun. The development of agriculture about ten thousand years ago was essential to the emergence of civilization. Success depended on knowledge of the seasons, weather for planting, and a bountiful harvest. Crop failure could mean famine and death. Humans, searching for comfort and affirmation, created idols of the Sun who could be worshipped to induce favorable outcomes. Earlier, we mentioned Helios and Sol, the Greek and Roman sun gods, but the inclusion of a sun god in polytheistic systems of belief was prevalent all around the world. At least 130 sun gods were created and worshipped, each with its own complicated story. Myths were made up and enriched, temples built, taxes levied, statues and images created, rituals developed, and sacrifices offered, usually animal but occasionally human.

The Egyptian sun god, Aten, was the deity of the first primarily monotheistic religion. About 1350 BCE, the Egyptian pharaoh, Akhenaten, and his wife, Nefertiti, became uncompromising monotheists declaring Aten to be the supreme state god. Akhenaten abolished other gods and had their images destroyed and their temples abandoned. This near monotheism was short-lived, and his successor, Tutankhamen, restored the offended gods by 1325 BCE in order to recover from an economic downturn, and he ordered their temples and images to be repaired.

Acceptance of official doctrine about the Sun was often mandatory. The Greek philosopher, Anaxagoras proposed in 440 BCE that the Sun was not a chariot pulled by Helios, but was instead a flaming ball. For this heretical assertion, he was initially sentenced to death, but later merely exiled. Galileo suffered a similar fate two thousand years later in 1632, when his proposal that Earth revolves around the Sun was declared to be heresy by the Roman Inquisition.

Nevertheless, no god, pharaoh, or king has yet to influence the Sun, and we now have a better grasp of its place in the universe. Our Sun and its solar system were formed 4.6 billion years ago, or about 9 billion years after the Big Bang. The Sun is by far the largest object in our solar system, accounting for 99.86 percent of the total mass in the entire solar system. More than one million Earths would fit conveniently inside the Sun. The gravity resulting from the mass of the Sun enables the most distant object in the solar system to be about 100 thousand AU away.

Our Sun is relatively young as stars go, evidenced by its small but significant content of heavier elements. (Recall that at the time of the Big Bang, hydrogen and helium were virtually the only elements yet formed.) As for all main sequence stars, the Sun is mostly hydrogen (73 percent) and helium (25 percent) with trace amounts of oxygen, carbon, iron, neon, nitrogen, silicon, and others. Notably, about six ten-billionths of the Sun is gold, which seems tiny except that it would more than fill all of the oceans on Earth.

The Sun consists of gas and plasma in several layers which fuses hydrogen into helium in its core thereby generating huge amounts of energy, which is radiated into space in various forms of electromagnetism and particles.

Plasma is superheated matter which is so hot (ten thousand degrees Centigrade or more) that electrons are released from their atoms, forming a charged, electrically conducting medium. Plasma is one of the four fundamental states of matter – the others being solid, liquid, and gas. Plasma is actually the most abundant state of matter in the universe.

Note in particular that there is no solid component of the Sun. It is all gas and plasma. The Sun's radius is 432 thousand miles, and since it is not solid, the Sun is able to rotate at differing periods at its equator (26 days) than at its poles (34 days). However, the Sun is almost a perfect sphere, with the difference of only 6.2 miles between its equatorial and polar diameters.

The core temperature is about 15.7 million degrees centigrade, and almost all of the energy produced by the Sun originates in the core within about 25 percent of the radius. The Sun currently fuses about 600 million tons of hydrogen every second into about 596 million tons of helium, converting the difference of four million tons per second into energy according to $e = mc^2$. This is enough to deliver abundant power to Earth from 93 million miles away. However, energy generated in the core takes at least one hundred thousand years to reach the surface of the Sun. Thus, although energy takes only about eight minutes to travel from the surface of the Sun to Earth, it was generated in the Sun's core around the time that Homo sapiens first appeared on Earth.

Although the Sun sustains us, it can also endanger us and threaten our power grid and electronic systems by hurling high levels of radiation and charged particles at us. You probably need no reminder to avoid looking directly at the Sun without risking permanent damage to your vision, or that spending too much time basking in sunlight increases the risk of skin cancer. One particle that you do not need to worry about is the neutrino, a neutral and almost massless subatomic particle produced during fusion. Only discovered in 1956, the neutrino is extremely difficult to detect because it has very little interaction with matter. A point of interest of neutrinos is that about 100 trillion of them pass through your body every second.

The Sun has a strong magnetic field which produces solar wind, a continuous stream of charged particles including electrons, protons, and alpha particles. (An alpha particle is an ion of an element such as oxygen or carbon). The interaction of the solar wind with charged particles in the Earth's atmosphere is the cause of the Aurora Borealis (the northern lights) and the Aurora Australis (the southern lights), which are the magnificent atmospheric light shows that occur near the North and South Poles. The force of the solar wind causes the tails of comets to always point away from the Sun. The solar wind travels at a velocity of about 500 miles per second and carries sufficient force that it would cause Earth's atmosphere to dissipate into space were it not for Earth's own magnetic field. (This is another one of the lucky breaks that has permitted life on Earth).

We do need to worry about the effects of the Sun's magnetic field, because any variation in a magnetic field can induce excessive electric currents in unprotected systems, and those induced currents can damage or destroy that system. This is a significant hazard to the 3000 satellites in orbit and potentially to astronauts. For very large solar magnetic events, there is also a hazard to systems on the ground.

The Sun's magnetic field can vary widely, often associated with sunspots, which appear darker than the surrounding areas, because of their lower temperature. Sunspots are often the site of strong variations in the Sun's magnetic field. The number of sunspots observed varies from year-to-year, in a cycle of about eleven years, and the observed annual number has varied from 13 to 157. The reason for this to be of great interest is that sunspots are indicators of intense solar events, including solar flares, which are unusually bright flashes of light, and coronal mass ejections (CME), large releases of several billion tons of plasma from the solar corona, the upper atmosphere of the Sun. CME's are potentially much more damaging than solar flares. Although CME's occur about once every five days, they are highly directional, and most do not affect our planet.

The largest CME impacting Earth occurred in 1859 and is known as "the Carrington event," because it was observed and recorded by a British astronomer of that name. The main effect was to cause auroras seen around the world. Telegraph systems in Europe and North America failed. Less severe CME's occurred in 1921 and 1960, accompanied by widespread radio disruption. It has been estimated that if the same event occurred today, the cost of the damage done to satellites and other modern systems could exceed two trillion dollars and take years to repair. There was a very near miss as recently as 2012, and some analyses calculate a significant probability of a Carrington-class event striking Earth within a few decades.

By coincidence, I am writing this chapter on June 1, 2021. On May 28, 2021, satellites detected a large-scale CME ejected from the Sun. We appear to have lucked out again, and the forecast is that Earth will be struck only a glancing blow, and only minor effects will be felt as the CME "sideswipes" our planet. However, this experience does provide some immediacy to what I have written above and to what you are reading today.

Whatever happens with CME's today, and other solar events in the future, about five billion years from now, the Sun will exhaust the hydrogen in its core, fusion will cease, and the core will collapse as a result of the loss of pressure which accompanies

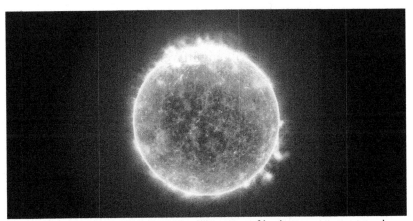

The Sun consumes about 600 million tons of hydrogen every second providing abundant energy to meet the Earth's needs. The Sun will run out of fuel in about five billion years.

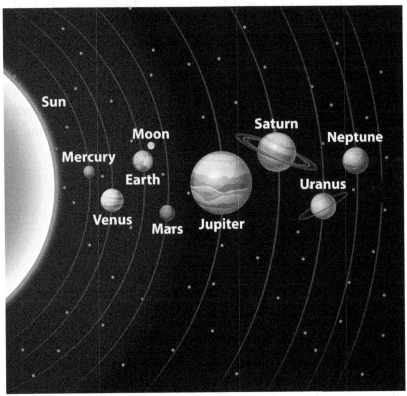

The Sun contains over 99.5 percent of the mass in our solar system, and its volume is more than one million times the size of Earth. The solar system includes eight planets, over 200 moons, and millions of asteroids and comets.

fusion. The resulting release of gravitational energy will cause the Sun to expand into a red giant, in a process requiring perhaps a billion years. The Sun will engulf Venus and Mercury and will come close to the present orbit of Earth. At the same time that the outer layers of the Sun are increasing, the core will continue to collapse inward leading to an increase in temperature from its present 15 million degrees to about 100 million degrees. At this temperature, the helium in the core will fuse into heavier elements, primarily carbon. During the red giant stage about one-third of the Sun's mass will be lost in forming a nebula. Following this, there will be several cycles with helium fusing into heavier elements which are lost to space as further nebular residue. These cycles will go on for about 100 million years. The Sun will become increasingly unstable with about half of its mass lost. The nebular remains of our Sun will be cast into space supporting the potential formation of new stars. Finally, the Sun will become a stable white dwarf star, about the same size as Earth, which will shine for perhaps a few trillion years before finally extinguishing its light forever and becoming an inert black dwarf.

Chapter 8
Earth - Your Home

———— ∞∞∞ ————

Every one you love, everyone you know, everyone you ever
heard of, every human being who ever was, lived out their lives
. . . on a mote of dust suspended in a sunbeam.
Carl Sagan
(1934 – 1996)

To see the earth as we now see it, small and beautiful
in that eternal silence where it floats, is to see ourselves as
riders on the earth together, brothers on that bright
loveliness in the unending night.
Archibald MacLeish
(1892 – 1982)

At long last, we are home to planet Earth. We find ourselves today 13.7 billion years after the Big Bang, 26 thousand light years from the center of the Milky Way, and 93 million miles from the Sun. Earth is one of eight planets which, together with Mercury, Venus, Mars, Jupiter, Saturn, Uranus, and Neptune, comprise our solar system. When it was formed about 4.5 billion years ago, Earth consisted mostly of molten magma and was completely inhospitable to life. It took another half billion more years before it could entertain even the simplest life and then another three billion years or so for more complex life forms to evolve. Homo sapiens did not appear until only a quarter of a million years ago – about 0.002 percent of the time since the Big Bang.

The Sun accounts for more than 99.5 percent of the mass in our solar system. The remaining mass forms the planets, moons, asteroids, comets, and other debris, as part of the same process that creates stars. As a newborn star begins to capture massive amounts of hydrogen and helium, it begins to rotate. This creates momentum for all of the accumulating mass, with the force of gravity being sufficient to hold the lighter elements in place, but insufficient to retain the heavier elements which spin off into the surrounding space. You can envision a wet dog shaking water off its fur as an instructive process, or even better, a salad spinner drying lettuce.

This heavier matter flies into space, in the general shape of a disc perpendicular to the axis of rotation of the star. Such a disc is called a "circumstellar disc" and is typical for solar systems throughout the universe. (We previously explained why stars and planets are spheres and why galaxies are often spirals.) As time passes, gravity enables local accumulations of mass with their own gravity, and which can attract further mass through a process called "accretion." For objects with enough mass, gravity will cause them to become spherical in shape. Their path through space then becomes elliptical in shape, some rotating around the star. We call these objects "planets." Other spherical objects stabilize via gravity into orbits around the planets, and we call these objects "moons." Smaller objects, usually not massive enough to be spheres, orbit either around the Sun or the planets, and are called "asteroids" or "comets." An asteroid is a small, rocky body orbiting the Sun, and a comet is a small object consisting of ice and dust. Altogether, our Sun's solar system includes eight planets, over 200 moons; more than one million asteroids larger than one kilometer in diameter, and millions of smaller ones; and about five thousand comets known, and as many as one trillion additional comets in the outer solar system.

The main significance of asteroids and comets to life on Earth is that they have erratic orbits around the Sun, and can collide with Earth with potentially catastrophic results. An estimated 30 million pounds of space debris enter Earth's atmosphere every year. A single asteroid with an estimated diameter of only seven miles caused the extinction of the dinosaurs about 70 million years ago. This was very bad news for the dinosaurs, who all perished, but very good news for the mammals, who survived this catastrophe. This ultimately allowed Homo sapiens to become the dominant species on Earth. The on-going threat is significant enough that NASA maintains a program to identify and monitor potentially catastrophic collisions, in the hope that we have adequate explosive technology to deflect an incoming object and avoid another dinosaur scale extinction event.

During the period when the solar system was forming, space near the Sun was occupied by trillions of pieces of debris, which collided frequently as planets were formed. Our moon most probably formed about 4.5 billion years ago, as the result of a giant collision of Earth with a small planet called Theia, which was about the size of Mars. This collision produced enough heat to cover Earth with magma oceans. Most of the debris was captured by Earth's gravity, and then collected in orbit around Earth to form the moon. The impact of Theia was a glancing blow, which caused Earth to spin on its axis much more rapidly than before. It is estimated that a day on Earth initially was only four hours long, and the moon and Earth were only about 15 thousand miles apart. Today, they are about 230 thousand miles apart.

Because of the close gravitational interaction between Earth and moon, Earth's rotation time is becoming slightly longer, and the distance from Earth to the moon is increasing slowly with each passing year. Sixty million years after the moon was formed, the day was 10 hours long. By the time life first appeared on Earth, about 4 billion years ago, the day was still just twelve hours long. When humans appeared, the day was about 24 hours long and we think of it as being constant over our lifetimes. However, the length of the day continues to increase at the rate of about 2 milliseconds per century, and the distance to the moon increases by about 12 feet per century.

Like the asteroid collision that wiped out the dinosaurs, the collision with Theia created side effects to humanity's benefit. Because of its significant mass, the moon enables the angle of the rotational axis of Earth to remain stable. Without this stability, the angle of the axis would vary much more, because of the changing gravitational forces of objects further away. Earth would wobble more around its axis, creating widely varying temperatures as it orbits the Sun. Humans might still exist, but much less comfortably. Perhaps even more importantly, the moon's gravity causes the daily occurrence of high and low tides of the oceans, and this may have been a crucial factor in enabling life to evolve from the sea to the land.

Earth is about the same age as the Sun. We estimate the age of the Sun as 4.6 billion years with the formation of Earth completed only 100 million years later, or 4.5 billion years ago. Even though our home was in place, it was far from ready for occupancy by humans. There was nothing to eat, no air to breathe, no oceans filled with water, and it was unbearably hot. Another 4.4 billion years would need to pass for Earth to become habitable by multicellular life.

The composition of the Sun is about 95 percent hydrogen and helium. The composition of Earth is radically different being 32 percent iron, 30 percent oxygen, 15 percent silicon, 14 percent magnesium, three percent sulfur, two percent nickel, 1.5 percent calcium, 1.4 percent aluminum, with trace amounts of the other 85 elements amounting to about one percent. Even though Earth is much smaller than the Sun, it is about four times as dense in mass per cubic foot because of the prevalence of the heavier elements. In fact, Earth is the densest of all of the solar system's planets and moons.

As it completed the accretion of its mass, Earth became structured into four distinct layers, not including the atmosphere: the inner core, the outer core, the mantle, and the crust. Each layer impacts life on the surface of the planet. The total radius of Earth is 3960 miles, with the inner core radius of about 800 miles, outer core thickness of about 1400 miles, mantle thickness of about 1750 miles, and the crust only 6 miles thick (beneath the oceans) to 40 miles thick (beneath the continents).

The inner core is a solid sphere consisting primarily of an alloy of iron and nickel along with other trace elements. Because of the intense pressure at the center of Earth, the temperature of the inner core is about 5700 degrees Centigrade, which is about the same as the surface of the Sun. The heat emanating from the inner core helps to stabilize the surface temperature of the planet.

The outer core has approximately the same composition as the inner core, but it is a fluid rather than a solid, due to the pressure of the outer core being lower than at the center of Earth. Because of the magnetic properties of iron, the relatively high temperature (about 4500 degrees Centigrade), and the rotation of the Earth, the outer core generates Earth's magnetic field. This results in magnetic poles located near our planet's true North and South poles. The magnetic field is essential to the ability of Earth to sustain life, because it protects from the potentially fatal radiation from the Sun. Without it, the Sun's solar wind would cause Earth's atmosphere to dissipate into space. Mars, for example, has no magnetic field, which is the main reason that it cannot sustain an atmosphere. Heat from the outer core is also an important contributor to maintaining a stable temperature on the surface.

The mantle consists primarily of the elements not heavy enough to sink into the inner or outer core. It largely contains chemical compounds of oxygen, silicon, magnesium, iron, aluminum, calcium, sodium, and potassium, which are most common in the formation of various types of rocks. The mantle includes almost 70 percent of Earth's mass and about 84 percent of its volume. Because most of Earth resides in its mantle, it is generally characterized as a rocky planet. Mars and Venus are also characterized as rocky planets, as distinct from Saturn and Jupiter which are called "gas giants."

Earth's mantle is cooler than the core, with temperatures ranging from 200 degrees Centigrade at the upper boundary to about four thousand degrees Centigrade at the boundary with the outer core. This is high enough that much of the mantle consists of liquid magma, including that which appears in volcanic eruptions. Diamonds and other jewels are formed within the mantle, but in much lower volumes than lava.

The mantle also includes a variety of radioactive elements that decay over time, thereby generating heat. These radioactive elements, primarily uranium and thorium, are estimated to yield up to half of the heat generated internal to Earth, although this is an area of on-going research.

After accretion of Earth was complete, the magma near the surface cooled and formed a crust, which covers the entire surface of the planet. The crust is the thinnest layer with an average depth of up to 40 miles below dry land, and depth of about 6 miles below the oceans. Like the mantle, it also primarily consists of a variety of rocks and minerals.

The upper part of the mantle, together with the crust, forms a fixed shell called the lithosphere, which has fractured into sections called tectonic plates. The lithosphere can be envisioned to be floating on the mantle. As a consequence of all of the forces involved, the tectonic plates are moving slowly in relationship to each other. They move at about the same rate that your fingernails grow. The plates evolved over time to form one giant supercontinent, called Pangaea, which began to break up about 200 million years ago into today's configuration of continents. Now there are seven major tectonic plates and several minor ones. The major plates are the African, Antarctic, Eurasian, Indo-Australian, North American, South American, and Pacific. It is easy to see on a map, for example, that South America and Africa would fit well if pushed together. Matching fossils and rock layers verify that they were once together. Two hundred million years from now, a map of Earth will look totally different than it does today.

The location where two plates meet is called a plate boundary. These boundaries are typically where earthquakes and volcanoes occur, and where topographic features such as mountain ranges and oceanic trenches are born. Plate boundaries do not usually coincide with the boundaries of oceans and continents. For example, the Pacific plate is entirely oceanic while the North American Plate includes all of the continent and much of the Atlantic ocean.

There are three types of plate boundaries: two intersecting plates might be diverging, converging, or sliding across each other. If two plates are diverging, the space between them is filled with molten rock from the mantle forming new crust, which slowly cools and solidifies while also possibly creating a chain of submarine volcanoes. If two plates are colliding, one may slide beneath the other in a process called "subduction" while exerting upward force on the other, causing beaches and mountains. The subducting plate also allows new molten magma from the mantle to rise to the surface, further enhancing the development of mountain ranges and plateaus. Alternatively, two converging plates may simply collide, causing the edges to be compressed and uplifted, as for example in the creation of the Himalayas. Before this occurred, the region of the Himalayas was, in fact, far below sea level only 40 million years ago. Where two plates are sliding along each other's boundaries, large forces build up which eventually must be released in the form of earthquakes, such as the great San Francisco earthquake of 1906.

Another crucial development in creating a habitable planet was the appearance of large quantities of water. At the initial formation of the planet, there were no oceans, and the surface was entirely the heated magma of the mantle. As the densest matter was pulled toward the center of the Earth, volcanic eruptions released vapor forming a gaseous atmosphere. There is no consensus as to exactly how the accumulation of water occurred. One theory is that, as Earth cooled below water's boiling point, water vapor in the atmosphere condensed in the form of continuous rainfall for millions of years. It is also thought possible that significant amounts of water came from asteroids and comets colliding with Earth, as part of the accretion process. Yet another theory is that Earth gained most of its water from the collision with Theia that caused the formation of the moon. Whichever theory or combination is correct, the oceans were fully formed by about 4.0 billion years ago, or only half a billion years or so after the planet was formed. Today, water is a defining feature of our planet, covering over 70 percent of the surface, and 97 percent of all Earth's water is in the oceans. If Earth were a perfect sphere, the entire surface would be water with an average depth of 1.5 miles.

Up until about 4 billion years ago, Earth was completely lifeless. If you were somehow able to live forward through time from the Big Bang, you might not have foreseen the creation of life, and you certainly could not have foreseen Mozart or Michelangelo. Although an atmosphere had formed, mainly from gases due to volcanic eruptions, it consisted primarily of a toxic mix of hydrogen sulfide, methane, and carbon dioxide. The surface was under constant bombardment from asteroids and comets remaining from the collision with Theia and from the formation of the Sun. Although there were oceans of water, there were also oceans of magma, and thousands of volcanoes heated from the Earth's interior. It would not have been a pleasant place to live, and geologists refer to this period as the Hadean Era, i.e., the Hellish Era.

Nevertheless, the planet had reached the stage when it could begin to support life, and it would soon do so. As we will see in examining the evolution of life, our planet has continued to evolve. Nothing about Earth remains constant. It changes at its own pace and in its own way including the length of the day, the climate, the brightness of the Sun, the distance to the Moon, the depth of the ocean, the polarity of the magnetic field, and the forms of life that it will support. We are on a voyage together

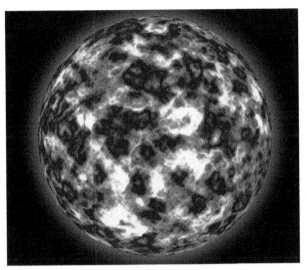

Four billion years ago, Earth was lifeless with a toxic atmosphere, largely covered with oceans of magma (molten rock) and under constant bombardment by asteroids and comets. This was called the Hadean ("Hellish") Era.

Modern astronauts are the first to see Earth from space. Frank Borman said "Maybe this really is one world and why the hell can't we learn to live together like decent people."

Our home planet today is beyond question one of the most beautiful places in the universe.

with Earth through time and space, and we are totally dependent on the environment it provides, with air, water, land and sunlight.

Perhaps the most remarkable characteristic of our Earth today, however, is its amazing beauty. There are simply no words in any language that can adequately describe this wonder, and so I will not try. Here is another You Tube suggestion for you: search for "Earth beauty" and choose a couple of hours of what you find there. Allow yourself time to sit back and take in the views, and consider your great good fortune to be here.

Chapter 9
Life on Earth - From One Cell to the Next

---⊛⊛⊛---

One of the deep mysteries of the origin of life is the almost indecent haste with which it arose on Earth.
Stephen Webb
(1961 – 2016)

It is not the strongest of the species that survives nor the most intelligent that survives. It is the one that is the most adaptable to change.
Charles Darwin
(1809 – 1882)

Although Earth was a hellish place, all the prerequisites for supporting life were in place only ten billion years after the Big Bang, or 4 billion years ago. The planet was in a very nearly circular orbit around the Sun, so the surface temperature remained within a modest range year around, and the planet was firmly in what is called the "habitable zone," or "Goldilocks zone." Water could exist in liquid form, and there was an abundant supply, ultimately constituting 60 percent of our bodies.

Equally important, there was an ample supply of four elements crucial to life: hydrogen, oxygen, carbon, and nitrogen. These elements include the chemical characteristics enabling them to form highly complex and stable molecules. Combinations of such molecules are called organic compounds. Millions of these organic compounds are known to science. Notably, two are ribonucleic acid (RNA) and deoxyribonucleic acid (DNA). These carry all of the information needed for the development, functioning, growth and reproduction of all living organisms. Initial life on Earth may have depended only on RNA, with DNA evolving later. Details of this are not known.

The crucial characteristic of living organisms is their ability to reproduce themselves. At some unknown place in Earth's ocean about four billion years ago, a single cell came into existence, with the ability to reproduce. This was enabled by improbable conditions of temperature, pressure, and electric charge (perhaps a lightning strike or within a volcano), and thus life began! You might like to compare the significance of this instant to the instant of the Big Bang and the instant your first cell was created in your mother. They are all important to you.

(There is also the alternative and not disproven theory that the first cell of life was created extraterrestrial and arrived on Earth by collision with an asteroid or comet. This only moves the instant of creation elsewhere and does not change the discussion of the further development of life on Earth)

We now have a single living cell that reproduces itself prodigiously and extends its reach over the planet. The reproduction process is not perfect, however, so each succeeding cell is just a tiny bit different than its parent. Some are defective and unable to reproduce and die without further repetition. Others are able to recreate more cells, some even better suited to their environment, and so more like them live and become more abundant. Over time, the characteristics of our single cell life become more varied than the original. This process of natural selection is how life first evolved and how it has continued to evolve over the ensuing four billion years. Grasping this notion is what earned Charles Darwin his place in history.

All living organisms are called either prokarya or eukarya. Prokarya are single-celled life forms, and do not have a distinct nucleus, nor fully formed genetic material. Prokarya still exist in abundance today in the form of bacteria. The original single cell organisms eventually evolved into multiple-celled organisms called Eukarya. This step took more than a billion years. Eukarya include all multicellular organisms, with a fully formed cell nucleus and DNA. Groups of similar organisms that can interbreed constitute a species.

Viruses merit special mention since they cause so much havoc on occasion in the form of pandemics, such as the one we are currently experiencing. A virus is a microorganism smaller than a bacterium (and thus is not a prokaryote) which cannot grow or reproduce on its own, but replicates only inside the living cells of an organism. There is not universal agreement on whether a virus can be described as a living entity, but with most arguing that since viruses cannot reproduce on their own, they do not count as a life form. Unfortunately, antibiotics that can successfully attack bacteria-causing disease are ineffective against viruses, since viruses do not attack human cells, but replicate from inside them.

The evolution of one species into another can occur in a variety of ways. When an organism reproduces, the offspring are inevitably slightly different than their parents. If reproduction is bisexual, then the offspring share the DNA of each parent, and thus are not the same as either. Also, the reproductive cells of each parent might have failed to exactly duplicate, because of random errors in the chemical process, or the cell might have suffered mutation due to external radiation. For all of these reasons, an offspring will have some different characteristics than its parents.

Those different characteristics may affect the chance of the offspring to survive and reproduce. Usually mutations produce negative, even life-threatening, effects, but occasionally the differences will be favorable. The giraffe with a slightly longer neck will be able to reach more foliage in tall trees. Perhaps a zebra can run just a little faster than average to escape a cheetah. Perhaps a cheetah can run just a little faster to catch a zebra. Perhaps a male peacock's plumage is just a little more attractive to potential mates. If an ice age is coming, and there have been at least five, those who are best able to tolerate the cold are more likely to survive. Over a long period of time, these favorable traits will become the norm for that species. Perhaps over a very long period of time, enough changes will accumulate to form an entirely new species which could not have interbred with the earlier generations.

Another way evolution can occur is if groups of a species become separated from each other due to migration or sudden changes in geography. Over time the two groups will change in enough ways that they can no longer interbreed. A new species will have been created. This can, at least in part, be caused over time by random changes in the DNA which defines some characteristic. This process is called genetic drift.

Humans also now influence how evolution occurs. Selective breeding of crops, such as wheat, corn, and rice, and domesticated animals, such as cows, chickens, sheep, and dogs, is an evolutionary technique contributed by our species over the last ten thousand years. Contemporary science now enables us to modify DNA to create new species. This is a capability that many find troubling.

All living creatures are passengers on Earth as it moves through space, and whenever this vehicle hits a bump, a lot of passengers can die because of the impact. These occasions are called "extinction events", and there have been five major such events (and about twenty minor ones) in the time since multi-cellular life first appeared. Their approximate time, percentage of species which became extinct, and the suspected causes are these:

1. **440 million years ago:** Up to 70 percent of all living species extinguished due to climate change associated with continental drift. By this time, larger aquatic life forms and even some land species had evolved.

2. **375 million years ago:** Up to 70 percent of all living species extinguished due to lack of oxygen in the oceans, cooling air temperatures, and volcanic eruptions.

3. **250 million years ago:** Up to 96 percent of all living species extinguished probably caused by asteroid

collisions, volcanic activity and the resulting climate change. This is the largest of the extinctions and is called "The Great Dying."

4. **200 million years ago:** More than 50 percent of all living species extinguished due to volcanic activity with resulting climate change and changing levels of the oceans.

5. **65 million years ago:** About 75 percent of all living species extinguished due to a major asteroid impact and subsequent climate change. This is the best-known extinction event because it extinguished the dinosaurs and paved the way for our own evolution.

One estimate is that about five billion multi-cellular species have existed on Earth, and well over 99 percent of them are extinct. There are presently perhaps 10 million living species, but we have identified and catalogued only about one million of them.

Biologists have devised a system called taxonomy to classify all organisms and to trace their evolution. All living things share the following attributes: they include one or more cells and are capable of reproducing themselves; they absorb energy in some form to support their growth, and they maintain a stable internal environment called homeostasis. There are nine levels of classification; Life, Domain, Kingdom, Phylum, Class, Order, Family, Genus, and finally Species. These categories provide a convenient way for us to trace and understand the principal phases of evolution of Homo sapiens.

The outline given below is a reasonable approximation of what many scholars believe about the path of human evolution. It should be noted, however, that there is not universal agreement about many of the classifications, facts, and dates given below.

Life – Living: We are among the descendants of single cell organisms (prokarya) which first came to exist between 3.8 and 4.2 billion years ago.

Domain – Eukarya: Eukarya are the first multicellular organisms which include a cell nucleus and DNA. Sexual reproduction was common. This is also the domain of many microorganisms such as fungi and algae. Eukarya first appeared about 2.7 billion years ago. It took over a billion years to advance from single cell organisms.

Kingdom – Animalia: An animal is characterized by voluntary movement (unlike plants), the ability to acquire and digest food, and to respond to the environment. Included are sponges, worms, insects, fish, reptiles, and many others. Marine life first appeared in the oceans about 800 million years ago. It took about 1.9 billion years to advance from eukarya.

Phylum – Chordate: Chordates have five key characteristics including a flexible chord along the back, a tubular nerve chord, gills, a tail, and a thyroid gland. Some chordates included a backbone. The oldest known fossil chordate appeared 500 million years ago.

Class – Mammal: Mammals include all animals that nourish their young with milk. Other principal characteristics include the presence of hair or fur, a region in the brain that specializes in seeing and hearing, specialized teeth, and a four-chambered heart. Mammals first appeared about 200 million years ago.

Order – Primate: Primates have large brains and increased reliance on visual acuity at the expense of sense of smell. Most primates have tails, except for apes. Most primates also have opposable thumbs. This group includes lemurs, monkeys, apes, and humans. Most primates sleep in trees at night. The first known primates appeared 65 million years ago.

Family – Hominid: The hominid group includes all modern and extinct species of humans, chimpanzees, gorillas, and orangutans (which comprise the Great Apes) plus their immediate ancestors. The first known hominid fossil is dated about 6 million years ago.

Genus – Homo: The genus homo includes modern humans (Homo sapiens) and our close relatives including H. habilis, H. erectus, H. Neanderthalensis, H. Denisovans and others. Main characteristics include a relatively large brain, a smaller and flatter face, smaller jaws and teeth, and increased reliance on tools. All members of this genus except Homo sapiens are extinct, although Homo sapiens interbred with at least two others (Neanderthalensis and Denisovans). The first known example was Homo habilis about 2.8 million years ago which descended from the hominid Australopithecus. Homo erectus was the first to leave Africa about 1.5 million years ago.

Species – Sapiens: At long last, here we are! The earliest Homo sapiens skeletons date from about 250 thousand years ago and were found in Western Africa.

Even though we have not catalogued all existing species, it is possible to estimate the total mass of all life on the planet, called the "biomass." Not surprisingly, plant life constitutes over 80 percent of the total biomass since there are more than three trillion trees. However, you might be surprised to learn that the

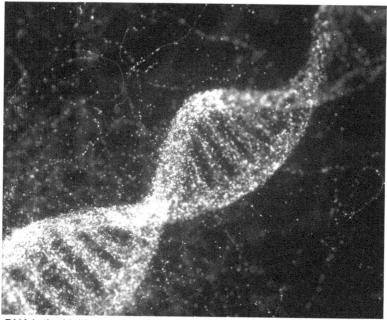

DNA is the highly complex molecule which encodes all of the information needed to define life. If the DNA in all 40 trillion cells in your body could be uncoiled and laid end to end, they would reach the Sun and back several hundred times, a total of about 30 billion miles.

prokaryotes (single cell life including bacteria) weigh about 17 percent of the total and that all animal life constitutes less than 0.5 percent of the total biomass. Human beings account for only about 0.01 percent of the biomass. In fact, we are outweighed by our livestock (e.g., cows and chickens) by 50 percent and our livestock outweighs all wildlife by a factor of ten or so. It is perhaps a bit humbling to realize that humanity is outweighed by bacteria by a factor of more than a thousand.

Our impact is far greater than our weight, however. We are probably in the midst of a sixth major extinction event with humans being its sole cause because of our pollution of the planet's oceans and atmosphere, because of our destruction of habitat, and because of our propensity to kill off other creatures for food and for sport. It is estimated that, since human civilization began, about 80 percent of wild mammals and marine mammals, 50 percent of plants, and 15 percent of fish species have perished.

Charles Darwin (Cambridge University, 1831) was the first to conceive evolution of life by natural selection. This is arguably the most significant idea anyone has ever had.

Rosalind Franklin (Cambridge University, 1945) was the first to suggest the structure of the DNA molecule. This is usually regarded as the greatest discovery in biology.

The current rate of extinction of species is much more than would otherwise be expected, and up to one million additional species face extinction within the next few decades. This rate of extinction is larger than any of the previous extinction events, and there is no guarantee that we will survive it.

Nonetheless, life has evolved and survived, and Earth has finally produced a good home for Homo sapiens, at least for now. It is worth noting the factors which converged to enable us to exist, and possibly to ponder the enormous good fortune of it all. If any of these were not true, we would not be here.

1. The rules of quantum physics which arose from the Big Bang support our existence. Physicists have envisioned many other potential sets of rules which would not have done so.

2. Supernova explosions created most of the elements which make up our bodies and our planet. We are indeed "star stuff," and our rocky planet provides what we need.

3. Our planet is in "the Goldilocks zone" relative to our Sun. That is, water can exist in liquid form rather than only steam or ice, and the temperature is not too hot or cold.

4. Our planet has a stable composition and an atmosphere that sustains us with oxygen. This oxygen is provided for us by the development of photosynthesis by cyanobacteria earlier in the Earth's evolution.

5. Our planet has a magnetic field which protects us from harmful radiation, and which helps to hold our atmosphere in place.

6. The diversity of life on Earth provides a food chain which sustains all species. Rocky soil became organic soil as microbes populated the planet, enabling agriculture.

7. The moon is a factor in creating the length of our day, stabilizing our axis of rotation, and causing tides which perhaps enabled evolution of life from the ocean to the land.

8. Our Sun has been relatively stable and has not bombarded us with radiation we could not survive.

9. The Ozone layer provides another buffer against harmful radiation.

10. There is abundant water, which sustains all life and composes much of it.

11. The gas-giant planet Jupiter serves as a "vacuum cleaner" which prevents asteroids and comets from reaching and colliding with the inner-planets.

12. Our planet is located in a relatively uncrowded part of the Milky Way galaxy, which reduces the danger of being sucked into a black hole or colliding with a star.

13. The Milky Way galaxy is not in danger of colliding with another Galaxy for a very long time.

14. If the dinosaurs had not been eliminated by an asteroid colliding with Earth, mammals would almost certainly not have evolved into us.

Even if each of these factors had a probability as high as one chance in a hundred (which is almost certainly too high an estimate), then the probability of all of them occurring would be 10^{-28}, an incredibly small chance. Thus, we are the result of a fantastic long-shot!

Two uncertainties confronting humanity going forward are whether or not this good fortune will continue, and whether or not we can overcome the menace created by our own success.

Chapter 10
Homo Sapiens - The Organism

─────⊗⊗⊗─────

What a piece of work is a man, how noble in reason,
how infinite in faculties, in forming and moving how express
and admirable, in action how like an angel, in apprehension,
how like a god.
William Shakespeare
(1564 – 1616)

A human body is a conversation going on,
both within cells and between the cells, and they're
telling each other to grow and to die.
W. Daniel Hillis
(1925 –)

Finally, 13.7 billion years after the Big Bang and as recently as 250 thousand years ago, the universe achieved the evolution of men and women, just like you and me, who biologically could be our brothers and sisters. This chapter concludes Part 2 by describing Homo sapiens, the organism, only from the viewpoints of our organic properties and construction. With this, we end our conversation about "comprehending your universe." The behavior of our species in the ensuing hundreds of thousands of years is the topic of Part 3. That is what is meant by "comprehending your world."

In the technical description of ourselves, we start at the level of atoms. An average sized human is made of about 7×10^{27} atoms (another one of those numbers too big to think about) consisting mostly of the four major elements of organic compounds; oxygen (65 percent of mass and 24 percent of atoms), carbon (18.5 percent of mass and 12 percent of atoms), hydrogen (9.5 percent of mass and 62 percent of atoms), and nitrogen (3.2 percent of mass and one percent of atoms). Altogether, these four elements comprise over 98 percent of our weight and over 99 percent of our atoms. The other one or two percent consists mostly of calcium, phosphorous, sulfur, sodium, chlorine, and yet smaller amounts of a dozen or so trace elements, all of which are essential to your health and well-being. Every element in your body originated soon after the Big Bang or in an exploding star.

All organisms consist entirely of a collection of cells. It takes about 40 trillion of about 200 different types of cells to make a human being. Each cell is enclosed in a porous membrane that allows oxygen and nutrients from food, such as glucose, to enter and waste products, carbon dioxide and water, to exit. The shape of the membrane is determined by the function of the cell. For example, nerve cells are very much longer than blood cells. Each cell contains a double helical strand of DNA, enclosed in its nucleus. The DNA is tightly coiled in order to minimize the volume it occupies. The cell also includes a strand of mitochondrial DNA, that converts chemical energy from food and oxygen into forms that the cell can use. RNA is also present in every cell to enable protein utilization, and to transmit genetic information from the DNA. The cell is filled with a gelatinous fluid, called cytoplasm, that suspends all of the cell's parts, called organelles. These perform all of the specific functions of the individual cells.

The DNA is a fantastically complex molecule, unique to every individual, containing all of the information necessary to define, build, and maintain us. The two strands of the helix are interconnected by about three billion base pairs which compose 46 chromosomes, 23 from each parent. (A base pair is two chemical bases bonded together, forming a rung of the DNA ladder.) These chromosomes define our heredity over all time and uniquely define us as an individual. If it were possible to uncoil the DNA, each strand would be a thin thread about six feet long. Thus, the total length of DNA in all 40 trillion cells of your body would be over 500 times the distance from Earth to the Sun.

Each chromosome includes a varying number of sections, called genes, that define one or more specific characteristics of the organism. Genes are the basic unit of heredity. Every person has a unique collection of about 22,000 genes, called the genome, that comprise the complete definition of how to build that person.

As one species evolves into another, many genes are shared and continue on through time. Thus, we can assess how much common heredity different species have. For example, Homo sapiens and chimpanzees share over 99 percent of their genomes. We also share about 70 percent of our genome with some species of worms. So, chimps are our brothers in evolution and worms are our cousins. We are all living things!

DNA must, of course, be able to replicate itself to enable growth from conception and replacement of damaged or dead cells. The double helix is the fundamental structure enabling cell replication. When replication is initiated, the two strands of the helix separate, and each strand serves as a template from which a new companion strand is formed. This enables each of the three billion base pairs to be replicated, in order to produce exactly the same genetic information in the two resulting DNA cells.

The human organism consists of about 200 types of cells with over 80 percent being red blood cells. Also, your brain is made of about 100 billion each of neurons and non-neuron cells called glia. Muscle cells constitute only one-thousandth of a percent of your total cell count, but those are the ones you go to the gym to strengthen.

The three types of cells most involved in the creation of a new human being are the sperm cell, that exists only in males, the egg, that exists only in females, and the stem cell, that is capable of forming itself into any of the remaining cell types.

Sperm cells are among the smallest in the human body, but are very numerous. A healthy male produces as many as 100 million new sperm cells per day, each one unique. Sperm cells can live in the female reproductive system for as long as five days, which gives them time to search for a female egg. The cell has a head, well designed to penetrate the egg, and a tail that it uses to swim upstream as fast as it can. Small as they are, they still carry half the genetic material needed to define a human. Of the 23 chromosomes, one of them, called the Y-chromosome, determines the sex of the baby, and the female's egg has no part in this determination.

On the other hand, the female's egg is the largest cell in her body, large enough to be seen by the naked eye. This size makes a good target for the sperm cell and the egg is designed to help the sperm penetrate. Every female carries a lifetime supply of up to one million eggs at her birth, all of them unique, and she releases one or two at every menstrual cycle, in a process called ovulation. Unless a sperm cell penetrates, the egg has a very short life cycle and will die within a day or so.

Although sex is determined only by the male and the vast majority of human DNA is an even mix from both parents, a small amount of DNA, called mitochondrial DNA (mtDNA), is determined only by the female. Mitochondria are structures within each cell, that help convert energy into forms the cell can use. The source of the mitochondria in the child is its mother's egg, and the male plays no role in this.

This fact makes it possible via DNA analysis to trace maternal lineage, since you inherited your mtDNA only from your mother, and she only from hers, and so on. If a woman has only sons, her mtDNA will not pass down to future generations. This also leads to the fact that at some time in the past, there was a "Mitochondrial Eve" from whom all living humans inherited their mtDNA. Mitochondrial Eve lived about 150 thousand years ago in Southern Africa.

When the joyful event of a sperm cell penetrating an egg occurs, they form a single cell of a fully defined human being. This cell, called a zygote, immediately begins to produce approximately one hundred identical copies of itself for the next four days while it travels to the uterus. All of these cells have the potential to become any of the 200 different cell types, and they are called stem cells. A newly developing human is called an embryo until the ninth week after conception, when it is called a fetus. By then, all of its major organs will have been formed. Finally, about 40 weeks after conception, a brand-new human makes the risky trip to its first breath of fresh air, beginning its journey through the universe and the world.

Almost as soon as we are born, our cells begin to die to be replaced with new ones. Cells die for several reasons, the most common being a normal cycle of reproduction and replacement defined by DNA. This process is called apoptosis. Cells may be damaged by radiation or by toxins and become unable to function. Mutations in DNA can occur in cell reproduction creating cells that become either unable to copy themselves further or to make defective copies at an accelerated rate possibly causing cancer. Many cell types have built-in limits to the number of times they can reproduce themselves, which partly explains why we age. Dead cells on the surface of the body or intestines are sloughed off. Others are absorbed internally or flushed out by the kidneys and liver. Altogether about one million cells in your body die every second and are replaced.

Each cell type has a natural life span ranging from a few hours for some white blood cells, to a few weeks for skin cells, to a lifetime for some cells in the nervous system, brain and eyes. Cells lining the stomach and intestines last only a few days because they live in the hostile environment of stomach acid and other corrosives. Your skin will be completely replaced about a thousand times during your lifetime. Overall, your body almost completely replaces itself with new cells every seven to ten years.

Since one view of Homo sapiens is as a giant collection of cells, we need to mention the roles of various types of microbes, including bacteria, viruses, and fungi. We live in understandable fear of these for good reason, since they kill tens of millions of people around the world every year. Diseases caused by bacteria include bubonic plague, leprosy, tuberculosis, syphilis, cholera, anthrax, staphylococcus, and many more. Viral diseases include flu (including Spanish flu and COVID-19), the common cold, herpes, polio, hepatitis, and many more. Diseases caused by fungi seem less cataclysmic, but include athlete's foot, and various other infections.

It may be a bit unnerving to know that there are more microbes dwelling in and on our bodies than there are cells made by our own DNA. Microbes live throughout the body, but primarily on the skin, in the mouth and digestive tract, in the intestines and the vagina. There are more than one thousand varieties of microbes that live in the stomach and intestines, and another thousand that live on the skin. In total, we all carry three to five pounds of microbes which number in the trillions. One estimate is that the average person has about 40 trillion human cells and 40 trillion bacteria and an even larger number of viruses. Recent studies have isolated viruses throughout the body including in the brain, blood, kidneys, and liver.

The good news is that most of the microbes are working for us, not against us. They help digest our food, extract vitamins and nutrients, and provide protective mechanisms that enable our immune system to recognize pathogens and defend against them. A newborn baby acquires its first accumulation of microbes during its passage down the birth canal, thereby providing initial stimulus to its immune system. The evolution of Homo sapiens has included not only its own cells, but the evolution of our on-board microbes, and we could not live without them.

The two hundred cell types and all of the accompanying microbes organize themselves into building all of the parts of your body. The study of human anatomy is organized into twelve basic body parts, such as skeleton and muscles, and over seventy organs, including heart and lungs. In order to comprehend your world, we do not need to go further in the study of human anatomy.

We have now reached the end of Part 2, Comprehending Your Universe. Only 250 thousand years ago out of the 13.7 billion since the Big Bang, a fully evolved Homo sapiens, just like you and me, stood naked on the threshold of our journey from Africa to the moon and beyond. Before we begin Part 3, here is a quick summary of the evolved characteristics of our species that enable us to to dominate and shape the world.

The most powerful tool in our evolved arsenal is our brain, which is at least three times as large as any other primate. The brain has the most complex structure of any known, and is capable of storing and recalling vast amounts of information, in addition to controlling all of our life functions. Equally important is our unmatched ability to imagine future events, and to plan what we need to do to cause those events to become reality. For early humans, that involved gathering and hunting for food, finding a safe place to sleep, and perhaps keeping a fire going. Survival was an everyday issue, and long-term retirement planning was not a worry. Average life expectancy over most of human history was about twenty-five years.

The capability to imagine would have been useless without the physical capability to turn imagination into execution. Happily, we are fully bipedal, thus freeing up our hands and arms for things other than swinging among the tree branches. Like many other apes, our thumb is opposable to our fingers, but in addition our fingers have become longer and straighter, so it is easier for us to grasp things, such as stone weapons, more firmly. In addition, our shoulders have become well suited to throwing things, such as rocks or spears.

In order to find sustenance, humans had to chase down, catch, and kill prey. Our legs became long and flexible, so we were good runners. In addition, we shed our fur in favor of a lighter layer of hair, and we developed an abundance of sweat glands, so we could stay cool while running long distances. In fact, although we are not nearly as fast over short distances as

Although humans come in many sizes, colors, shapes, and talents, we all share 99.9 percent of our DNA in common. We are one species, and everyone is beautiful in their own way!.

cheetahs, for example, we are the best endurance runners on the planet. Therefore, as a group, we could chase anything down, but first we needed to organize and cooperate.

Organizing to chase down food would be more successful if we could communicate with each other. The shape of our mouth, larynx, and tongue evolved in such a way that, unlike any other creature, we could make recognizable and repeatable sounds, which ultimately became words and language. The tongue is one of our toughest muscles, although not entirely because of its use in speech.

With speech, humans began to talk in groups about accomplishing common goals, such as surviving. We could imagine what we needed to do and make plans; we could make primitive weapons to defend and feed ourselves; and together we could chase anything down or avoid anything chasing us. Taking all of these attributes together means we could work together to survive and prosper. We were on our way to civilization, and it would only take another few hundred thousand years until we began to live together in large groups which would be called cities, states, nations, and empires.

For our species to survive, we also needed to go forth and multiply, and evolution has made us very good at that as well. Our brains have about sixty thousand thoughts per day, and for most adult males, some of those thoughts are about sex and participating in it. That is probably somewhat less true for females, but nevertheless they can participate at any time, conceive once a month, and deliver a child once a year or so. Now there are 7.9 billion of us.

Delivering a newborn human was a dangerous event for both mother and child. Mortality with childbirth has been common over most of human history. The infant mortality rate over history is estimated at about 27 percent. One reason for this is that as the size of our brain and skull increased, it was more difficult, painful and dangerous for the baby to travel down the mother's birth canal. Even so, the infant's brain is only about one quarter of its ultimate size, and so an infant human is totally helpless at birth and requires intense nurturing through childhood and adolescence. This was one reason why the human family group evolved, since support of the mother is essential to the survival and development of the child, and the support by the father is essential to the survival of both.

All of these evolved attributes of Homo sapiens remain in place today, although in many cases their application is radically different. J. K. Rowling devised Harry Potter from her imagination and used her well-evolved hand to write the stories, which have become the best-selling books ever. We no longer have to run long distances to catch dinner, but Pheidippides ran 26 miles from Marathon to inform Athens of the victory over the Persians in 490 BCE, and the first sub-two-hour marathon race was run in 2019 by Eliud Kipchoge of Kenya. Baseball greats Nolan Ryan and Sandy Koufax had the good fortune to use their arm and shoulder to throw no-hitters rather than spears.

The first modern human lived about 250,000 years ago, but the first civilizations appeared in Mesopotamia only 6,000 years ago. Thus, over 95 percent of the time that Homo sapiens has existed was pre-civilization, and much of our modern behavior is still strongly influenced by the earlier times. In the balance of the book, we will try to comprehend how it is going.

Part 3
Comprehending Your World

Chapter 11
Preview of Part 3

---⎯∞∞⎯---

In all civilizations we've studied, all cultures that we know of across the Earth and across time have invested some kind of attempt to understanding where they come from and where they are going.
Neil deGrasse Tyson
(1968 –)

"Natural" man is always there, under the changeable historical man. We call him and he comes – a little sleepy, benumbed without his lost form of instinctive hunter, but, after all, still alive. Natural man is first prehistoric man – the hunter.
Jose Ortega y Gasset
(1883 – 1955)

We have traced the development of the universe from the Big Bang to the landmark of the fully evolved Homo sapiens species, that is you and me. The emergence of science and the scientific method over many fields of study has allowed us to understand underlying principles of our universe and our existence, subject to the development of better knowledge. Before humans existed, there was no question of good versus bad, or right versus wrong. Also, science does not address the question of why things happened outside the framework of objective reality. It does not consider why the Big Bang occurred; why the laws of gravity are as they are; or why natural selection governs the evolution of life. It claims to illuminate WHAT things are, not WHY things are. The WHY is still in the realm of philosophers, theologians, politicians and dreamers to work out as best they can. However, if these luminaries take positions contrary to what has been verified by science, they are probably incorrect, no matter how persuasive or authoritative.

Part 3, is fundamentally different in that it reports what humans have accomplished. The human mind has enabled many civilizations, in all of their dimensions, to be created. However as in Part 2, we maintain the objective of comprehension, not judgment. Following is an example which illustrates why I have chosen not to attempt judgment on past events.

Humanity populated all of the habitable parts of Earth by 14,000 years ago, when we first reached the tip of South America. The path and sequence of migration began in Africa, proceeded north through Europe and east through Asia, across the land bridge to North America and then south through South America. Civilizations developed independently in every region, but Europe, the Middle East, and China had about a four thousand year head start on the American Continents.

As the scientific revolution began about 1500, several European states (notably Spain, France, England, Portugal, and the Netherlands) developed both the technologies and the thirst for expansion to the West. They had sailing ships, navigation skills, and weapons to enable them to go wherever they wanted and subdue anyone they encountered. Initially, their primary motivation was to develop trade routes to the Far East, but as exploration of the Americas developed, their motivations quickly evolved to colonizing and extracting wealth.

By this time, the Americas already had a population in excess of 50 million. There were a significant number of well-developed agricultural civilizations in Central and South America including Olmecs, Mayas, Aztecs, and Incas. In North America, there were about six hundred native tribes, each typically with its own language and customs. In comparison, the total population of Europe in 1500 was about 75 million, so the populations of Europe and the Americas were similar.

Columbus and other European explorers appeared, and then everything changed. European civilizations found a home for those citizens who wished to emigrate to a new and more hopeful land, ultimately leading to the formation of the countries of the Americas. On the other hand, by 1600 the population of the previous occupants of the Americas was reduced by about 90 percent, with an estimated total death toll of around 50 million due to disease and slaughter. For Europeans, the Americas represented hope and wealth. For those already here, it was an alien invasion, leading to subjugation or annihilation. There were eight times as many deaths resulting from colonization of the Americas than the Holocaust, and this enabled European civilizations to replace native cultures.

So, what should we think of this result? It is not my goal to try to make that judgment. Rather, the goal in Part 3 is to summarize the human experience so that we better comprehend it, not judge it. In general, I take the view that things were not good or bad – they just were. I leave judgment to you if you care to make it.

Chapter 12
Hunter/Gatherers - Down From the Trees

———∞∞∞———

In prehistoric times, Homo sapiens was deeply endangered.
Early humans were less fleet of foot, with fewer natural
weapons, and less well-honed senses than all the predators
that threatened them. Moreover, they were hampered in their
movements by the need to protect their uniquely immature
young – juicy meals for any hungry beast.
Robert Winston
(1940 –)

For 2.5 million years humans fed themselves by gathering
plants and hunting animals that lived and bred without their
intervention. . . . Why do anything else when your lifestyle
feeds you amply and supports a rich world of social structures,
religious beliefs and political dynamics?
Yuval Noah Harari
(1976 –)

O ur species, Homo sapiens, (genus "Homo", species "sapiens") was given its name in 1758 by Swedish scientist, Carl Linnaeus, with the literal Latin translation "wise human being." This occurred almost a century before Charles Darwin proposed evolution by natural selection. In retrospect, Homo loquacibus (talkative human) or Homo caedes (murderous human) might be more accurate. At the time, we were the only identified member of the genus "Homo", but subsequently we know that there were about twenty others, called hominids, all of which are now extinct.

Before we discuss the life of the first Homo sapiens, we should pay tribute to what was achieved by earlier hominids (humans) since we share many of their characteristics. The first hominids appeared about 2.8 million years ago, and the earliest named species are Homo habilis and Homo rudolfensis. These humans walked bipedally and invented the first stone tools. Humans also first became able to control fire around two million years ago.

There is evidence that Homo erectus learned to cook food about 900,000 years ago. Some theorize that the emergence of cooked meat enriched the human diet, thus promoting the development of the larger cranial capacity and smaller teeth which characterize Homo sapiens.

Homo sapiens was not the first hominid species to migrate out of Africa. That honor belongs to Homo erectus who spread throughout Eurasia beginning about two million years ago. Other early humans also migrated throughout Europe, Asia, Australia, and the Pacific islands by about 500,000 years ago. Two notable cases are Homo neanderthalensis ("Neanderthal man) in Northern Europe and the Asian Denisovans, both of which later inter-bred with Homo sapiens, and thus form a small part of our DNA.

One measure of success for a species is how long it survives. The oldest living creature more complex than a microorganism is the sponge which has existed for almost 800 million years. Dinosaurs patrolled Earth for about 165 million years. The longest periods of survival among hominids are Homo erectus who lived for about 1.8 million years and Homo habilis who survived for about 1.3 million years. From that perspective, Homo sapiens is a late arrival, even among hominids, having existed for less than 300 thousand years so far.

The time interval between humans first use of simple tools made from stone until the first development of agriculture is called the Paleolithic ("old stone") Era. The Paleolithic Era lasted from about 2.5 million years ago until about 10,000 years ago. Estimates of the total human population at the end of the Paleolithic Era range from one to ten million.

In reciting theories and dates such as the above, it is well to remember that the uncertainties in dates are large, and the theories are based on inference from limited data. The fossil remains of only about six thousand humans have been unearthed. Paleontologists have become extremely skilled at interpreting fragmentary fossils and other evidence to make large-scale inferences about ancient species. Much can be learned from a skull or a jawbone or a pelvis. Moreover, new evidence is acquired with every decade and every newly-discovered fossil site. Two new Homo species have been identified as recently as 2021. Of course, each new discovery is accompanied by lively debate about its veracity and impact on earlier research. This is the scientific method at work.

The broad outline appears to be stable, however. The earliest appearance of the genus Homo occurred about 3 million years ago; Homo sapiens appeared less than 300,000 years ago; the first agriculture-based civilizations occurred at most 10,000 years ago followed almost immediately by the first empires leading to the modern era of industrialization which began only 500 years ago. Thus, the time period of civilizations is only about three percent of the time since Homo sapiens appeared and less than one percent of the time since the first human. It is reasonable to posit that many of our current instincts, behaviors and characteristics were defined long before the first book was written.

All life on Earth is inherently competitive and violent. In order to reproduce ourselves, we must compete to find food and the most suitable mate. In whatever group we find ourselves, we compete for the role most advantageous and attainable to us, from emperor on down. All life fits somewhere in the food chain and is either predator or prey or both. Homo sapiens has often found it advantageous to kill our own kind in pursuit of personal or group survival, passion or aspiration.

For early humans, acquiring sustenance was a daily priority. Lacking tools and hunting skills, diet consisted of fruits, seeds, nuts, insects and grasses with only limited access to meat by hunting weaker game or scavenging. Initially, man was more likely prey than predator in the face of crocodiles, large birds, leopards, sabertooth tigers, bears, snakes, and perhaps other species of early humans. Our modern instincts and behavior derive in part from this period in our evolution.

Hominids turned the tables as early as two million years ago, becoming the most successful predator on Earth, an "apex predator." By 600,000 years ago, we had developed spears and knives plus the ability to collaborate in groups to chase down and surround almost any beast. By the end of the Paleolithic era, most large mammals were hunted to extinction including mammoths, mastodons, sabretooth tigers, oversized lions, giant sloths, North American camels and horses, and many others. We have continued this process up to modern times including the dodo bird (1681), the passenger pigeon (1914), and the northern white rhinoceros (pending – there are two females in captivity and no males as of 2021). As discussed earlier, Homo sapiens is the cause of today's on-going extinction event which has already wiped out about 80 percent of land mammals.

Take a moment and try to envision yourself as one of the first Homo sapiens, with the same mind, body, and capabilities you have today. Perhaps you are a young mother with a child or two. Your instincts have prepared you to defend them at any cost, and you live in a small band of perhaps twenty other persons to help you. You are able to make expressive sounds, but there are not yet many words in your mind. Assuming you live in a moderate climate in central Africa, you are unclothed and do not care. Your children followed sex with one or more males in your band, but you have no knowledge that there is cause and effect. Every day is a new adventure in locating enough to eat, either by gathering nuts and other plant life, or perhaps by helping to chase down whatever game your stone weapons permit. (Gathering was not strictly a female pursuit and hunting was not strictly a male activity.) You live in fear of the predators who would like to feast upon you and your children, and your band may have to defend itself against other hominids who share your area. You know that night follows day, and that by nightfall you had better be in your cave or hut, perhaps with a fire maintained to provide some warmth and scare off intruders. You might know how to cook the day's kill. Every so often, you and your band might need to move to a more promising place to sustain you. Overall, it is the life you know, and it seems worth living and defending. You have plenty of time to do things other than hunt or gather, and perhaps you start to wonder what will happen to you when you die, and how you can control the fates that might befall you. Except for the accident of the century of your birth, this could have been your life, and you would have tried hard to sustain it, just as you are trying to sustain your life today.

The brain's ability to reason is surely Homo sapiens' most important evolutionary feature. However, there is a strong case to make for our ability to speak as being next in line. We are the only species capable of forming words, developing language, and singing. Most paleontologists believe that speech capability probably did not begin until perhaps 150,000 years ago, although some argue that it may have come as early as two million years ago. Either way, it was decisive, because speech is what enables us to communicate whatever comes from our brain. It enables us to work in groups on whatever task is at hand. It enables us to articulate our ideas and to receive the ideas of others and to debate the differences. We can express our feelings ranging from threatening our enemies to wooing our lover. Speech was crucial in developing hierarchies and establishing who would lead and who would follow, ultimately resolving who would be king and who would be slave. We are highly social animals, and the socialization begins with speech.

Nobody knows what the first spoken word was, but it might have been some sound of alert, and the human vocabulary grew from there. Today, the total number of words in the dictionaries of the leading languages range from around half a million to a million. Most adults have vocabularies in the range of about

25,000 words. Speech remains the most important quality of leadership, with the lives of politicians and leaders of industry defined primarily by their schedule of appearances, meetings, conferences, and briefings.

Homo sapiens evolved in south-central Africa about 250,000 years ago and then, perhaps aided by primitive speech, spent most of the intervening time slowly populating Africa. It was only about 70,000 years ago that we first reached Europe, but after that the pace of our diaspora sped up with only 10,000 more years to span Asia, another 45,000 years to cross the Bering land bridge and then finally only another 4,000 years to reach the tip of South America. Roughly, this journey of about 15,000 miles took about 60,000 years or a quarter of a mile per year. Moving along was made necessary by the hunter/gatherer nomadic lifestyle, which required migration as demanded by climate change, food supply, environment and competition.

On average, bands of Homo sapiens who were 1,000 miles apart in distance were 4,000 years away from each other in time, and had no memory of each other's existence or history. Thus, every group had to solve all of its own new problems as best it could without direct reference to earlier experience. This accounts for all of the subsequent diversity which our species displays in terms of physical appearance, ethnic groups, cultural differences, and language. There are over 7,000 languages spoken in the world today and almost another 1,000 that are known to be extinct. Today, there are still over 5,000 distinctly identifiable ethnic groups.

Although we are all still one species, differences in physical appearance began to occur because of genetic drift among populations no longer in contact with each other and because of local adaptation to the environment. A good example is that as Homo sapiens moved into Northern Europe, skin color lightened because there was less need for melanin in the skin to protect from solar radiation and more need to extract Vitamin D from the Sun. The subsequent ramifications of this simple adaptation have been enormous.

We remain a remarkably genetically homogeneous species, with the DNA of any two humans differing by only about one part per thousand DNA base pairs. Among the implications of this are to confirm that modern humans originated in one place (central Africa) and that at one time our breeding population was as low as 10,000 persons. We are nearly brothers and sisters to everyone we encounter.

Given that there were about 20 species of hominids, it is curious that Homo sapiens is the only survivor. After all, there are many different kinds of cats, bears, and rodents, so why only one human? There are several different possibilities, but no definitive evidence. One is that our superior brain power made us more adaptable to the several changes in environment and natural catastrophes that occurred over the paleolithic era. Another is that as Homo sapiens populated Earth and encountered our fellow humans, we outcompeted them and drove them to extinction much as we probably did with mastodons and giant sloths. It is certain that we overlapped with Neanderthals in Europe for over a hundred thousand years before they died out leaving traces of their DNA in ours, proving that there was some interbreeding with Homo sapiens. A third possibility is simply that we were lucky to survive several near misses and other hominids were not. For example, there was a volcanic eruption about 70,000 years ago which may have reduced the Homo sapiens population to less than ten thousand and possibly eradicated other hominids. This event also explains why there is so little diversity in our DNA compared to most other living species. We will never know, but it is interesting to speculate about what civilization might have been like had there been multiple competing species of humans.

Although the business of having to hunt and gather your meals every day while avoiding being eaten was doubtless stressful, this probably took only six or seven hours a day. Thus, early Homo sapiens had time to use the creative potential of their minds to consider other matters. Perhaps the most important was what can be done to ensure survival, both during this life and hopefully after this life ends. By about 30,000 years ago, humanity was dispersed widely over the planet, and so each band or tribe had to address these matters independently, and thus different solutions in detail were reached. Many, however, followed similar paths in believing that there are independent spirits existing throughout nature, whether in rivers, trees, mountains, rocks, Earth, the moon, or the Sun. What immediately followed was to imagine how to influence all of those spirits to grant good outcomes, whether abundant food, surviving the tiger, or living forever in some afterlife. It was important to carefully define the characteristics of each spirit, especially deciding what pleases it and what makes it angry. This led to the development of rituals to entice the spirits to obey their bidding, including drawing pictures on the walls of caves, developing chants and dances (accompanied by music), and offering sacrifices such as food, animals or a young male from an adjacent tribe. Many groups identified specific individuals, called Shamans, to lead this effort and to decide what was right and what was wrong and how everyone should behave.

These earliest systems of belief are called Animism, and thousands of them in various forms were invented independently around the planet. Some of the spirits of Animism were called gods and became the basis of later polytheistic religions. It is remarkable that nearly every ethnic group developed belief in some external set of forces, spirits, and gods who made the rules of existence and who could be influenced by human exhortation.

Animism was closely involved with the introduction of the dimensions of civilization which we call culture. The earliest known painting appeared on the wall of a cave in Spain about 40 thousand years ago. The oldest known statuette was found in Germany and dates to 35 to 40 thousand years ago. The oldest known musical instrument was made by a Neanderthal and is 60 thousand years old. It is a flute with four holes carved from the thighbone of a bear. The first funerals and burials occurred at least 50 thousand years ago. There is evidence of early rituals associated with the python occurring in Africa 70 thousand years ago.

Eventually, Homo sapiens began to wear clothing, although there is no general agreement about when or where this first occurred. One plausible possibility is that mankind first donned fur pelts when the colder parts of northern Europe were first populated around 100 thousand years ago. The earliest evidence of needles used to fashion fur dates from more than 30 thousand years ago. Earliest clothing materials were most likely fur, leather, leaves, or grass. Some groups of hunter/gatherers never found the need for clothing, and there are still remote parts of Africa and South America where clothing is not normally worn.

As humans migrated around Earth, they were confronted with the problem of providing shelter from the elements and predators. The common vision is that early humans lived in caves, but most locations required other solutions, and only a small percentage lived in caves. The first shelters were made from stones and tree branches in the form of simple huts or tepees. They were often circular in shape, possibly with a hearth in the center. Other materials, including stone slabs, bones, and animal hides were used as available, and these were used in most shelters throughout the Paleolithic era. The earliest use of clay to form dried bricks did not occur until about 9,500 years ago, long after Homo sapiens had migrated around Earth.

The principles of evolution suggest that for any living creature, the most important goal is to reproduce so its DNA will survive into the future. That is as true for Homo sapiens as any other animal, so full comprehension requires that we understand how we reproduce ourselves. It should come as no surprise that our minds have combined with our evolution to produce a system unlike any other species. We have been very successful at this since there are now nearly eight billion of us.

The roles of the human male and female in reproduction are radically different. The male plays a direct role only at the instant that his sperm penetrates the egg. He can perform this role many times per month, and most males are eager, if not driven, to do so. Thus, from an evolutionary perspective, his goal is to spread his DNA as broadly as possible, using any attainable mate as his vessel. For the female, her investment is vastly greater. She produces one or two eggs per month for fertilization and only then until menopause. If she becomes impregnated, her life is altered first by enduring the pregnancy and then surviving childbirth. She then has a helpless baby who must be nurtured for a decade or two. This is a life-altering and life-threatening event. Therefore, again from a purely evolutionary perspective, her best strategy to propagate her DNA is to carefully choose the mate who presents the best DNA and the ability to support her and her offspring. Therefore, males are designed to be indiscriminate in choosing sex partners and females are designed to be selective. In our era, the average female lifespan is a few years longer than for males. For most of history, exactly the opposite was true, primarily because of the high mortality rate for females at childbirth.

As hunter/gatherers populated the planet and formed several thousand cultures, each had to define how to cope with the fundamental matter of the sexual behavior of its members. The variations are legion. Most (but not all) cultures are patriarchal, that is males are dominant, The variations are too broad to summarize here, except to note that an important part of comprehending any civilization is to understand how it attempts to manage sex. Just envision all of the current debate concerning what sorts of behavior (homosexual, bisexual, asexual, transgender, polygamy, polyandry, abortion, divorce, prostitution, pornography etc.) occur and how anxious politicians and clergy are to dictate what is permissible and what is not.

Humans are highly social animals who rely on working together to achieve common goals. Our intelligence and the development of language were crucial parts of becoming the dominant species on Earth.

Humans' ability to collaborate, plus our skill as long distance runners, transformed us from prey to apex predator. As a consequence, most large mammals were hunted to extinction.

Also note all of the recent episodes of leaders whose careers have been altered by revelations of their own conduct.

At the individual level, do not underestimate the importance of sex in most lives. After all, that is how we all came to be. One recent study suggests that males think about sex almost 20 times a day and women more than ten. Almost 40 percent of internet download activity is related to sex with males accounting for two-thirds of it and women for one-third. The variety of sexual acts that humans have contrived is astounding. Under the Wikipedia entry for "Sexual Acts," the number of categories of behavior totals one hundred and sixty-two.

One of the last vestiges of absolute slavery today is the involuntary trafficking of females, mostly young girls, across international borders for purposes of sexual exploitation. Up to five million women are enslaved in this manner to support a 150 billion dollar per year growth industry. It is hard to imagine a more shameful activity short of genocide.

We conclude this discussion of the Paleolithic Era by summarizing the state of the species as the era drew to a close about ten thousand years ago. Homo sapiens successfully populated every continent except Antarctica, and many islands. Our numbers grew from a few thousand to a few million. We developed sufficient tools, weapons, and techniques to advance from prey to Apex Predator, helping to eliminate all other hominid species and most large mammals. We split up into several thousand ethnicities with accompanying languages, customs, rules, and physical appearance. We began the search to understand our significance and place in the cosmos through the development of animist systems of belief. We started to develop the arts in the form of paintings, sculpture, music, and dance. Altogether, this is not a bad body of achievement for only a quarter million years. Homo sapiens now stood on the threshold of civilization. Next, we move on to attempt to comprehend what has happened in the most recent ten thousand years.

Chapter 13
Civilizations - From the Farm to the Empire

———— ❦ ————

Without agriculture it is not possible to have a city, stock market, banks, university, church or army. Agriculture is the foundation of civilization and any stable economy.
Allan Savory
(1935 –)

Every civilization carries the seeds of its own destruction, and the same cycle shows in them all.... The people invent their oppressor, and the oppressors serve the function for which they are invented.
Mark Twain
(1835 CE – 1910 CE)

Although the hunter/gatherer lifestyle enabled Homo sapiens to populate Earth, it had significant limitations. Population of individual groups or bands was a maximum of about 100 souls because of limited availability of food. It was often necessary to move from place to place due to seasonality, climate change, exhaustion of local food sources, or competition with other bands. This made it impossible to build permanent dwellings or to have many permanent possessions. The nomadic hunter also had the potential to become prey as well as predator.

Homo sapiens had survived the most recent ice age which peaked about 60,000 years ago with much of the planet enveloped in ice. The glaciers had receded by about 15 thousand years ago, leaving regions of year around moderate climate. (We are still in this period of moderate climate, although we may be in the process of ending it by our own uncontrolled heating of the atmosphere.) As luck would have it (and luck is the operative word), there were a few places on the planet which were conducive to year-round occupancy.

Another prerequisite for continuous occupancy and growth is an abundant and wholesome supply of water to sustain human life and provide a reliable source for crop growth and harvest. A third circumstance supporting year-round stability is a significant variety of edible plant life, including grains, nuts, berries, legumes, and tubers. Finally, there needs to be a locally available range of animals suitable for domestication. These creatures must be able to breed in captivity, mature rapidly, be relatively docile, and willingly conform to existing in groups or herds. Lions, bears, and cobras make poor farm animals, but horses, cattle, pigs, sheep, goats, chickens, camels, llamas, reindeer, and yaks all had wild ancestors, ultimately tamed, and selectively bred to meet human needs. Selective breeding of both plants and animals was the first human intervention in the processes of evolution by natural selection.

Four locations satisfied all of the required conditions and ultimately became the sites of the first civilizations. These were in Mesopotamia, supported by the Tigris and Euphrates rivers; Egypt, supported by the Nile river; the Indus River valley in present day India and Pakistan; and China, supported by the Yellow and Yangtze rivers. All initially developed independently of each other and each had its own specific collection of domesticable plants and animals.

In every case, the development of agriculture formed the foundation of civilization in two most essential ways. First, it enabled the population to be fed without requiring everyone to be involved in subsistence every day. Second, if food was relatively abundant, the population in a small area could grow far beyond the limitations of the hunter/gatherer culture. Taken together, many more people were available to think about and do other things than finding food. Successful agriculture enabled everything else, and if agriculture fails for any reason, then civilization fails quickly thereafter. (This is as true today as it was 10,000 years ago.)

The transition from hunter/gatherer culture to agrarian civilization was not simple. In fact, 5,000 years elapsed between the first permanent villages and the first Mesopotamian civilization, and another 5,000 years elapsed between the first agrarian civilization and the Industrial Revolution.

There are many skills required to progress from gathering a few stalks of wild grain to planting a productive crop of wheat covering many acres. These include learning how to condition the soil; developing tools to support planting and harvesting; understanding when to plant and when to harvest; developing irrigation systems; developing fertilizers; and developing means of storing the surplus. Each of these evolved over time and represented significant contributions to the ensuing civilizations. The process and time required to transform wild aurochs into a docile herd of cattle were similarly time-consuming, as were the domestication of other animals.

Since this chapter is about civilization, we should discuss more carefully what is meant by the term. There is no universally agreed definition, but this one will suffice: "A civilization is a complex society that is characterized by social stratification, a form of government, urban development, and symbolic systems of communication." The English words – civilization, city, and citizen derive from the corresponding Latin roots, "civilis, civis, and civitas." Every civilization is built around one or more cities that provide its nucleus and heart. A city is a consolidated location where many people gather to perform specialized roles and work together towards common goals. The largest city in the first civilization, Sumeria, was Uruk which reached a population of between 40 thousand and 80 thousand in about 2800 BCE. At the peak of the Roman Empire about 3000 years later, Rome may have reached about 450 thousand inhabitants. Today, another 2000 years later, the most populous city in the world is Tokyo, Japan with a population in 2021 over 37 million. Moreover, there are over 500 cities in the world with populations greater than a million. Compare that with the roughly 100 maximum humans for a single band of hunter/gatherers! Consider just how much food and water is required daily to support all of these urban dwellers and how much waste they produce.

The four oldest agrarian civilizations all developed independently at roughly the same time. Each made fundamental and unique contributions to the development of later civilizations.

The Sumerians (4100 BCE – 1750 BCE) of Mesopotamia (i.e., the Tigris and Euphrates Rivers) are generally regarded as the oldest of the four. Their contributions included writing, the first number system, sun-dried bricks, wheeled vehicles, irrigation, the plow, maps, and metallurgy. They are also credited with the first major contribution to literature in the form of the Gilgamesh Epic, which foretold the myth of the great flood, later adopted by monotheistic religions.

The Egyptians (3100 BCE – 30 BCE) of the Nile River con-tributed an independent writing system, the first mathematics and geometry, shipbuilding, surveying, astronomy, paper in the form of papyrus sheets, the inclined ramp, and mills for grinding grain.

The Indus Valley civilization (2600 BCE – 1700 BCE) of the Indus River developed the first system of standardized weights and measures and made important advances in metallurgy with copper, bronze, lead, and tin. They also invented an independent writing system which unfortunately has yet to be translated accurately.

The Chinese civilization (1600 BCE – 220 BCE) of the Yangtze and Yellow Rivers was geographically the most isolated, but nevertheless made a number of important basic contributions including the first paper and printing, gunpowder, the compass, the seismograph, and silk fabric. Like the others, the Chinese invented their own system of writing.

The dates given above are in every case arbitrary and approximate, and there was significant development in each area prior to the earliest dates given. Relative to the total time frame of Homo sapiens, these four civilizations can be regarded as developing virtually simultaneously and independently.

Much as human speech and language were the most important factors in the development of hunter/gatherer cultures, writing can be seen as the most important factor in the development of civilizations. Writing is simply speech in permanent form, which has the wonderful property of allowing minds to connect across both space and time. Writing is the main enabler of what is often called "collective learning." Writing is a great tool to communicate not only what is factual and novel, but it also enables us to immortalize whatever we can imagine. The telling of history, law, science, news, poetry, and fiction across the ages have all been made possible by the written word.

Every agrarian civilization independently organized its members into well-defined social hierarchies. At the very top were the rulers, whether called king, pharaoh, emperor, or caesar, and their close lieutenants, who exerted absolute authority over all. Every early civilization also adopted a polytheistic system of belief accompanied by a supporting class of full-time clerics whose mission was to create, interpret, and enforce the gods, rituals and doctrine. The ruler and the religious leaders maintained close contact, with the ruler declaring himself to be either ordained by the gods or descended from the gods or even to be a god himself and thus able to invoke "the divine right of kings", a concept which survived until the eighteenth century or later.

First and foremost, every civilization had to be able to defend itself and perhaps conquer neighboring cultures. This required a warrior class of men willing to die for the cause, or preferably able to kill for it. Next came the craftsmen and other skilled laborers directly responsible to execute and extend technology. These initially included such crafts as stone cutters, metal smiths, weavers, fishermen and perhaps twenty others. This is the level from which most of civilization's inventions and advancements came. Still, the vast majority of workers were farmers whose output sustained everyone else. Of course, about half of the inhabitants were female, whose treatment varied widely among civilizations although in most cases they were subordinated to males. In Egypt, for example, Cleopatra (69 BCE – 30 BCE) rose to be Pharoah. However, women were mostly chattel to men, and in some cases in India, were expected to immolate themselves on their husband's funeral pyre when he died.

Finally, and in every case of early civilizations, there were slaves. The main need for slaves arose from the necessity for energy to execute large-scale tasks, whether harvesting crops or building pyramids or the Parthenon. Prior to the Industrial Revolution, the cheapest energy source was human labor, perhaps amplified by beasts of burden. The class of slaves could be created as a punishment for misdeed, or by conquering adjacent civilizations and enslaving their members. Some had no choice but to become slaves to avoid starvation, and the practice of selling oneself, one's own children, or wife into slavery was common. Slavery was an accepted and even applauded practice. Aristotle said, "For that some should rule and others be ruled is a thing not only necessary, but expedient; from the hour of their birth, some are marked out for subjection, others for rule." At the height of its democracy, the average Athenian family owned three or four slaves.

In every case, the leaders of agrarian civilizations invested wealth and slave labor in developing four primary types of structures: palaces and tombs to provide domiciles appropriate to god-like kings in this life and the next; temples (including monumental statuary) to honor and placate the gods of the moment; fortresses to defend against invaders; and infrastructure, e.g., roads, dams, and aqueducts, to support growing populations. These spurred development of the technologies required for their successful execution as well as providing purpose and employment. Who can fail to be amazed and enthralled by the great pyramids and temples of Egypt; the Parthenon and other temples on the Greek Acropolis; the Roman forum, the Great

Wall of China, and the temples of Angkor Wat? The Colosseum in Rome was the largest amphitheater in the Roman Empire, seating 50 thousand. It witnessed the deaths of 400 thousand people and one million animals during its 400-year active use beginning in 72 CE. It is fascinating to realize that these were built two millennia or more ago and were then a part of day-to-day life. All were built without benefit of cranes, bulldozers, freight trains, or power drills. There are hundreds of such remnants all over the globe, leaving enormous legacy for us. Several Roman dams and aqueducts continue in use today (albeit with repair and improvement), and the great dome of the Roman Pantheon remains the largest unreinforced concrete dome in the world. You could make a rich life exploring these great artifacts, as have many archaeologists.

From very early in their development, agrarian societies recognized the benefits of trading with other cultures. The first long-distance trading occurred between the Mesopotamian and Indus Valley beginning in about 3000 BCE, a distance of about two thousand miles. The first major trade routes were the rivers which spawned the civilizations, i.e., the Nile, Indus, Yangtze, etc. Trade on the Mediterranean Sea was a primary impetus to the development of the Phoenician, Carthaginian, Athenian, and Roman cultures. The use of domesticated horses and camels enabled caravans to transport goods in significant quantities over long distances. Beginning in 130 BCE in China, a network of overland trading routes, called the Silk Roads, linked all of the regions of the Eurasian world. The Silk Roads were the most essential avenues of worldwide trade until supplanted by ocean-going ships in the fifteenth century. Spices, textiles, and precious metals were the principal early trading commodities. Silk from China was highly prized in Rome, and at one point silk was literally worth its weight in gold. The Chinese were able to keep secret the process of making silk for over a thousand years, and death was the penalty for revealing this process or attempting to smuggle silkworms out of China. Bartering was the original mechanism for making trades, but this was soon replaced by the invention of money, usually in the form of silver or other metallic coins.

In addition to enabling the sharing of material wealth, the trading routes also enabled the distribution of ideas, inventions, language, philosophy, culture, weapons, religions, and all of the other intangible aspects of civilizations. Agriculture provided the capability for the great growth in the number of humans, and trade provided the mechanism for the universal sharing of knowledge. Together, these two factors drove the enormous acceleration in human development which led to the Industrial Revolution.

Throughout modern times, the fundamental vehicle for the sharing of knowledge has been the written word. A recent estimate is that about 130 million different books have been written. This is an excellent example of an essential invention which is based upon contributions from several civilizations. The first writing was done by Sumerians on clay tablets in about 3200 BCE. The Egyptians invented papyrus in about 2900 BCE, a much better medium for conveying writing, followed by their invention of parchment in about 2500 BCE, which was even better. Alexander the Great inspired the great library at Alexandria, which housed an estimated 100 thousand volumes by 100 CE. Paper was first invented in China in about 100 CE and ultimately reached the Middle East about 500 years later. It took another 500 years for paper to become common in Europe. Of course, paper also requires ink, which was independently invented in Egypt, and China at about the same time. The Chinese invented the first movable type in about 1040 and thus had a long head start in paper books. They had the largest libraries for many centuries. Finally, movable type based on metal letters was invented by Johannes Gutenberg in Germany in about 1450, completing the last major step in producing books in large numbers for easy transport throughout the world. This invention occurred as the world was on the threshold of the Industrial Revolution and was one of its principal enablers.

A negative impact of the close contact among civilizations was the frequent outbreak of pandemics. Epidemics and pandemics have caused over a billion deaths from the beginning of agrarian civilizations, about the same number as in all wars. An epidemic is defined as a sudden increase in the number of cases of a disease, and an epidemic becomes a pandemic when the disease spreads over a large geographic area. Historically, pandemics are a side effect of the hunting and domestication of animals which can transmit diseases to humans.

The earliest known pandemic was smallpox, probably originating in Egypt in about 3000 BCE. It ultimately spread throughout north Africa, Europe, and Asia as a result of the trade activity among the civilizations along the Silk Roads. Ultimately, smallpox reached Japan where it killed about one-third of the population. Smallpox remained endemic throughout the world with a mortality rate of 30 percent, and is estimated to have caused nearly ten percent of all deaths before the widespread use of the first vaccine developed in 1796. Even so, It is estimated to have killed between 300 million and 500 million people in the 20th century alone before finally being eradicated via systematic vaccination around the world.

The most famous pandemic to date was the Black Death or Bubonic Plague which struck Europe in the fourteenth century causing up to 200 million deaths in just four years. European population decreased by 30 to 60 percent, and the mortality rate was up to 80 percent of those who contracted the disease. The disease originated in Asia and was possibly carried to Europe by flea-infested rats travelling via the Silk Roads or by trading vessels. Since there was no medical science to explain and defend against the plague, many radical and fanatic cures were attempted including astrology and begging God's forgiveness for imagined sins. In Germany, about 150 Jewish communities were liquidated in the belief that Jews had poisoned the water supplies, perhaps foretelling later events there.

There have been perhaps twenty other significant pandemics caused by cholera, measles, influenza, typhus, malaria, tuberculosis, HIV/AIDS, and, most recently, COVID-19. The Plague of Athens (430 BCE) devastated the Athenian population and killed its great leader, Pericles. This plague helped bring an end to the Golden Age of Greek democracy. The Antonine Plague of 165 CE, probably caused by smallpox or measles, wiped out about a third of Roman population and contributed to the downfall of the Western Roman Empire. The Plague of Justinian, also caused by bubonic plague, reduced European population by more than one-third in 550 CE. Prior to COVID-19 of 2019, the most significant recent pandemic was the Spanish Flu of 1918, which caused up to 100 million deaths, or up to five percent of the world population at the time.

Ultimately, the survivors of pandemics develop immunities (so-called "herd immunity") which spread throughout the populations of Europe and Asia via the trade routes. Unfortunately for them, the civilizations of the Americas had not developed any immunities to the Eurasian plagues, and they quickly succumbed when the Europeans arrived around 1500 (as we noted in the Preview to Part 3).

Although the present pandemic of COVID-19 is significant, so far only about 0.1 percent of the world's population has perished from it to date due to modern practices of quarantine and masks together with the rapid development and deployment of effective vaccines. Imagine the effects on our civilization of the rapid loss of half of our people were it not for modern medicine.

A curious characteristic of our species that, under enabling circumstances, charismatic leaders emerge who set the goal of conquering as many adjacent civilizations as possible for the purposes of converting, indoctrinating, governing, subjugating, enslaving, looting, and/or destroying them while purposefully causing the deaths of millions. Such accumulations of power are usually called "empires." Moreover, empires establish within their own populations the enthusiastic and unquestioning adoption of the leader's goals (e.g., "Hail Caesar" and "Heil Hitler"). Here is a short list of some of the more notable and successful emperors including the name of their empires and their lifespans.

Sargon of Akkad (???? – 2279 BCE) – Akkadian Empire
Thutmose III (1479 BCE – 1425 BCE) – Egyptian Empire
Cyrus the Great (580 BCE – 529 BCE) – Persian Empire
Alexander the Great (356 BCE – 323 BCE) – Macedonian Empire
Ashoka the Great (304 BCE – 232 BCE) – Indian Empire
Han Wudi or Emperor Wu (156 BCE – 87 BCE) – Han Dynasty
Julius Caesar (100 BCE – 44 BCE) – Roman Empire
Augustus Caesar (63 BCE – 14 CE) – Roman Empire
Attila the Hun (406 CE – 453 CE) – Barbarian Empire
Charlemagne (742 – 814) – Holy Roman Empire
Genghis Khan (1162 – 1227) – Mongol Empire
Tamerlane (1336 – 1405) – Extension of Mongol Empire
Napoleon Bonaparte (1769 – 1821) – French Empire
Adolf Hitler (1889 – 1945) – German Third Reich
Joseph Stalin (1878 – 1953) – Soviet Union

Note that this urge to conquer spans the entirety of human civilizations, from the first to the present. It is debatable whether the United States as presently constituted is an empire or not. However, Americans are fond of referring to its President as "the most powerful man in the world" and the United States maintains more overseas military bases than any nation in history. Altogether, it is possible to name and study about 150 distinct empires that have, with the possible exception of the United States, come and gone to be replaced by the current worldwide collection of nation-states.

Uruk, the largest city in Sumeria, had a population of about 40 thousand in 3100 BCE. World population was about 40 million.

Whether you choose to call it an empire or not, the United States of America (USA) has been the world's predominant civilization since the end of World War II. Its essential progeny can be stated in one very long sentence. The hominids (3 million YA – 50,000 YA) begat Homo sapiens (250,000 YA to present) who became the hunter/gathers who populated Earth and begat the first civilizations including the Sumerians (4500 BCE – 1900 BCE) and Egyptians (3150 BCE – 1650 BCE) who begat the Greeks (700 BCE – 323 BCE) who begat the Roman Empire (625 BCE – 475 CE) which begat the nation states of Europe (350 CE – present) which begat the British Empire (1600 – 1945) which begat the USA (1776 – present).

The founding declaration of the United States was a direct descendant of ideas developed in England. In 1776, Thomas Jefferson famously wrote: "We hold these truths to be self-evident, that all men are created equal, that they are endowed by their Creator with certain unalienable rights, that among these are Life, Liberty, and the pursuit of happiness." John Locke (1632 – 1704)

150

By 100 CE, Rome had become the largest city in the world with a population of 450 thousand. World population was approaching 200 million.

Tokyo, Japan is today's most populous city with 38 million residents, about the same as world population five thousand years ago. World population today is approaching eight billion, or two hundred times more than in 3000 BCE. Today there are more than five hundred cities with populations in excess of one million.

in England a century before had written: "All mankind being all equal and independent, no one ought to harm another in his life, liberty, or possessions." Locke's and Jefferson's declarations applied at the time only to male, Caucasian, and property-owning Christians. The implied exclusions resound into our lives today.

Altogether, since the first Sumerian civilization in 3100 BCE, about 150 distinct civilizations existed in the ensuing 5200 years, usually as a result of triumph in conflict. Thus, a new civilization begins on average every 40 years or so. Each develops its own culture and hierarchy and imagines itself to be divinely ordained and immortal. A civilization provides life, purpose, sustenance, training, protection and motivation to its individual members in return for their allegiance. It does wonderful things (art, architecture, literature, music, laws, knowledge, science, kindness, etc.), and terrible things (warfare, tyranny, slavery, persecution, imprisonment, torture, genocide, etc.), and leaves its mark. Perhaps the most compelling thing a civilization does, however, is to provide shelter for its citizens to go about their lives. They fulfill whatever role has been consigned to them whether high or low. They strive to make a living, reproduce and find satisfaction in their lives while being much more interested in what is for dinner than what is happening in the palace or temple. Then, after an average duration of only two to four hundred years, circumstances become adverse for every empire. Perhaps the climate changes, natural catastrophes may occur, or key resources are depleted. A pandemic may come to destroy many lives, or there may be civil discord and violence. A stronger adversary may conquer. Several of these may accumulate at once. Finally, every empire ceases to dominate and is replaced by a successor.

Chapter 14
Feudalism to Domination –
Europe Conquers the World

—— ≋ ——

*Colonialism subdues in many dulcet guises. It conquered under
the pretext of spreading Christianity, civilization, law and order,
to make the world safe for democracy.*
Wole Soyinka
(1934 –)

*Books are the carriers of civilization. Without books, history
is silent, literature dumb, science crippled, thought and
speculation at a standstill.*
Barbara Tuchman
(1912 – 1989)

In the millennium from 1000 to 2000, European civilization
achieved predominance. European nations conquered territory
on every inhabited continent, spreading their languages, religions,
technology, clothing, culture, and weapons. The Beijing Central
Philharmonic Orchestra's most recent recording included works
by European composers Mendelssohn, Saint-Saens, and Vivaldi.
One of the few globally unifying activities today is the Olympic
Games, which originated in ancient Greece. They restarted in
recent times in London, England, and feature competitions in
events mostly of European origin. Street scenes from around the
world show many people dressed nearly identically in Western-
style garb. On the darker side, the machine gun, the jet fighter,
the guided missile, and nuclear weapons are all products of
European minds, which have armed the world's militaries.

From the vantage point of 1000, this was not a likely outcome.
After the fall of the Western Roman Empire in 395 CE, Europe
came under assault by the Mongols from the east, the Vikings
from the north, and the Islamic Empire from the south via the
Iberian Peninsula. In Europe, there was no central governing

authority, and society partitioned itself into thousands of small units, each contending for survival. Thus, in the centuries immediately following Rome's collapse, Europe was engulfed in lengthy wars including the Hundred Years' War, between England and France (1337 to 1453) and the Eighty Years' War, between the Netherlands and Spain (1568 to 1648). Europe was also ravaged by the Black Death (bubonic plague pandemic) in the fourteenth century, which killed about 50 million people, amounting to 40 percent of the total population. The Little Ice Age, which was a period of unusually cold weather lasting for over a century, beginning around 1290, caused numerous famines across Europe in the thirteenth and fourteenth centuries.

In contrast, the Islamic Empire was in its Golden Age between the eighth and thirteenth centuries. Baghdad became the largest city in the world and was a center of culture, commerce, economics, and science. The great works of the ancient Athenians were translated into Arabic and preserved for posterity. Significant advances were made in mathematics, chemistry, and astronomy.

Similarly, China's golden age began about 600, and China regarded itself as the center of the universe with the world's oldest culture and society. Among its great inventions were silk fabric, paper, printing, gunpowder, and the compass. The Silk Road was the network of trade routes connecting China with the Middle East and Europe. China believed that it had far more to offer in trade than it needed in return. At the time of Columbus' voyages to the New World, China was building trading ships many times larger than the Europeans could manage. China elected to focus its trading efforts toward India rather than eastward, giving Europe free rein to colonize the New World. In the remainder of this chapter, we will trace the factors which led to Europe's ascendance.

Feudalism was the economic system prevailing in Europe from the fifth through fourteenth centuries. This system was strictly hierarchical, with kings at the top. They ruled under divine authority granted to them by the Catholic Church, already established as the state religion by the Romans. Land was the principal asset in this agrarian scheme, so conquering neighboring land was the optimum way to increase personal power and wealth. Each king granted the right to portions of his land to nobles who pledged their loyalty, military force, and taxes in return for the king's protection. Land was further partitioned to vassals (tenured landholders), who in turn pledged their loyalty, obedience and military service. The vassals' military commitment was typically 40 to 90 days per year. This could include the service of knights, who were also granted land in return for loyalty, service, and income. At the bottom of this hierarchy were peasants, or serfs, who were bound to the land for life and were essentially slaves who worked for only enough to feed themselves in return for protection from the hierarchy above them. The peasant class amounted to about 90 percent of the population.

In the event of dispute among kings or invasion by outsiders, the only option was to resort to violence. Private war among local lords was commonplace, resulting in more than 10,000 castles built at strategic points to provide refuge and defense. In modern Europe it often seems that every mountaintop and river junction features a medieval ruin as a tourist attraction.

There were many thousands more churches than castles though, and the Catholic Church claimed primacy of authority. The vast majority of the population were Catholics, and there was no other Christian denomination as of yet. The church was well-established, with the Pope in Rome having divine authority derived from Jesus and the Apostle, Peter. The hierarchy of Cardinals, Bishops, and Priests was in place. People often attended church several times a week. The Church paid no taxes and claimed tithes from all, and the Church and its clergy were wealthy as a result. The Church regulated life from birth to death, and beyond. Ghastly tortures were a common form of punishment and source of public entertainment, for sins such as heresy or witchcraft. The ultimate power of the Church was excommunication and thereby consignment to eternal damnation.

Clerics developed many transparent schemes to increase their wealth and power, including the sale of indulgences which, for a large fee, promised forgiveness for sins and salvation.

These were the features of feudalism in Western Europe: 1) constant fear of attack from neighboring fiefdoms and external forces, whether Vikings, Islam, or Mongols; 2) domination of daily life by the Church including heavy taxation and fear of excommunication; 3) nearly universal illiteracy (except for the clergy) accompanied by fear and suspicion; 4) loss of the writings and art of Greece and Rome; 5) a short average lifespan of about 35 years; and 6) near-slavery for the great majority of people. Small wonder that this period, lasting from about 500 to 1000, is often called the "Dark Ages."

However, in the late medieval era (1000 – 1500), a variety of loosely-related factors combined to end feudalism, transform Europe, and enable its domination of the world.

Perhaps counterintuitively, the Black Death, beginning in 1348, was a significant factor in the escape from feudalism. Depopulation reduced the size of the labor force, so survivors were more essential to the production of food, and thus commanded better treatment. Land, which was the primary source of wealth, diminished in value as there were far fewer people to occupy and use it. As it became clear that the Church was helpless in the face of plague, faith in the power of religion diminished. Some towns were so depleted that religious observances ceased altogether. The relative power of women increased, because with so many men dead, women were now able to own land or run a business. Those who survived had every reason to question the authority and primacy of both temporal and religious hierarchies. This questioning helped trigger two major movements commonly known as the Renaissance and the Reformation. These were essential in bridging the gap between the Dark Ages and modern times.

A second major factor bringing an end to feudalism was the consolidation of power in those kingdoms most successful in conquering rivals. This ultimately resulted in the formation of fewer and larger kingdoms, later to become nation-states. Most were governed by monarchs who pronounced themselves to be divinely authorized. The major monarchies became competitors for power and colonization of the Americas and Africa. The dates of origin and first monarch included these:

England – King Athelstan, 925 (England conquered Ireland in 1169, Wales in 1277, and Scotland in 1296)

Germany – King Otto I, 962

France – King Hugh Capet, 987

Portugal – King Afonso I, 1139

Spain – King Ferdinand of Aragon and Queen Isabella of Castile, 1469

Habsburg Empire –Emperor Maximilian I, 1493

Russia – Tsar Ivan IV (also known as Ivan the Terrible)**, 1547**

The Netherlands – Stadholder William I, 1581

Italy did not consolidate into a nation state until much later (1861), but the city-states of Venice, Florence, Milan, Naples and Rome were vitally important in late medieval times as centers of trade, banking, the arts, and science.

Much of the history of Europe for the last millennium centers around the struggles of these kingdoms against each other, culminating in the two world wars of the twentieth century.

A third major factor in the emergence from the Dark Ages was the invention of the movable-type printing press by Johannes Gutenberg around 1436 in Mainz, Germany. Printing presses had been independently invented and used extensively in China and Korea much earlier, but were unknown in Europe. The printing press enabled mass production of books and newspapers at low cost. Previously, books required laborious copying by hand. The cost of a single book often exceeded the cost of a house, and the largest European library in 1300 contained only 300 manuscripts. But by 1500, after only 50 years of printing, there were more than 20 million books in Europe. Literacy in much of

Europe increased from less than 10 percent in 1450 to over 60 percent by 1750.

It is interesting to compare the effects of the printing press in 1500 with those of the smartphone in 2000. They both increase the amount of information available to the common person, and they reduce the time to acquire it. They both can entertain, inform, educate, and influence. They both make it more difficult for authority to limit and censor, since what is said cannot be unsaid. Once they exist, it is hard to imagine life without them. In 2010, the fraction of people who used a smartphone was probably less than the literacy rate in 1450, but by 2020, most people depended on one. A significant difference between the two is that one can directly comment on anything delivered by the smartphone, whereas the printed word is less interactive.

Martin Luther said that "printing is the ultimate gift of God and the greatest one." Francis Bacon wrote in 1620 that the three inventions that forever changed the world were gunpowder, the compass and the printing press. (All three were first invented in China, but unknown in Europe for several hundred years.)

The ready availability of printed books facilitated two major developments of the late Middle Age, called the "Renaissance" (i.e., "rebirth") and the "Reformation" (sometimes called "the Protestant Reformation") which spanned the fourteenth through seventeenth centuries. Both of these involved questioning Catholic orthodoxy. The Renaissance produced the philosophy of Humanism which focuses on roles of reasoning and evidence, forming the basis of the modern scientific method. The Reformation, for its part, led to the founding of many separate denominations, generally identified as Protestants. These include Lutherans, Calvinists, Anglicans, Baptists, and others.

The Republic of Florence is generally regarded as the birthplace and center of the Renaissance. Florence had become a major trading center, in close contact with the trade routes to the Islamic Empire and China. As such, it had accumulated great wealth, as represented by the Medici family. Lorenzo de Medici (1449 – 1492), in particular, was a patron and sponsor of the arts.

Francesco Petrarca, "Petrarch," (1304 – 1374) laid the foundation for the Renaissance by reviving interest in the culture and works of the ancient Greeks and Romans. This led to the philosophy of humanism with a focus, not on religion, but on the inherent capabilities of individual humans. Together with the patronage and support of the wealth of Florence, an explosion of new ideas was created in art, architecture, and science.

Among the great Italian figures of the Renaissance were Dante Aligieri, Giovanni Boccaccio, Leonardo da Vinci, Michelangelo Buoarotti, Niccolo Machiavelli, Amerigo Vespucci, Sandro Botticelli, Donatello, Raphael, Giotto, Lorenzo Ghiberti, Pico della Mirandola, Galileo Galilei, and Nicolaus Copernicus. This is a roster whose impact on civilization is comparable to the giants from the Golden Age of Greece and those from England listed in Chapter 4. Leonardo da Vinci, in particular, is often regarded as the most diversely talented individual in history for his contributions to art, science, architecture, anatomy, music, mathematics, astronomy, and weaponry as described in his journals of more than 13,000 pages.

The seminal event of the Reformation was the posting, in 1517, by Martin Luther of his 95 theses, "The Disputation of the Power and Efficacy of Indulgences." Among his theses was "It is certain that when money clinks in the money chest, greed and avarice can be increased." In addition to opposing the sale of indulgences by the clergy as a money-making device, Luther believed that the Bible is the source of truth about God, not the Pope or Church doctrine. Luther translated and published the Bible in German, enabling members to read it in their own language, rather than depending on priests to recite it to them in Latin.

With this bold move, made possible by the printing press, Luther quickly became a celebrity all over Europe, and an inspiration to other religious reformers including John Calvin (founder of Calvinism) and Henry VIII (founder of the Church of England). Luther's translation of the Bible into German was a best-seller, with over 200 thousand being printed during his lifetime. His *On the Freedom of a Christian* and *A Treatise on Christian Liberty* both espoused the freedom of the individual in his relationship with God, independent of the Church hierarchy. Luther said, "A Christian man is the most free lord of all, subject to none."

Taken together, the Renaissance and the Reformation laid the foundation for modern concepts of democracy, human rights, independence of thought, individualism, humanism, and separation of church and state.

The Catholic Church did not accept this without firm resistance. Pope Leo condemned Luther's writings, and ordered Luther's works be burned, and that anyone who owned or read them would be excommunicated. Luther was declared to a heretic in 1521, four years after posting his 95 theses.

In response to the rise of the protestant movement, the Catholic Church issued a series of internal reforms and strongly resisted the spread of other denominations. This led to a period known as the Counter-Reformation, extending from the establishment of the Roman Inquisition in 1542 to the end of the Thirty Years' War in 1645.

The Inquisitions (Roman, Spanish, and Portuguese) were responsible for prosecuting individuals for crimes involving religious doctrine with the goal of maintaining religious orthodoxy. Among those prosecuted were Copernicus and Galileo for the "foolish and absurd" proposition that Earth moves around the Sun. Both Catholics and Protestants were fond of prosecuting women for witchcraft (exhibit the later Salem witch trials in Massachusetts), with an estimated total of 40,000 or more being executed between 1400 and 1800.

The struggle between the varying forces of the Reformation and Counter-Reformation subjected Europe to constant warfare between 1400 and 1700. More than fifty distinct wars occurred. Many involved conflicts of both religion and the struggle for power among kingdoms. Perhaps the most devastating was the Thirty Years' War (1618 – 1648) which killed at least thirty percent of Germany's population, a larger percentage than either World War. It began as a conflict between Protestants and Catholics, and like World War I, developed into a general war involving most of Europe, ultimately killing more than ten million people.

Martin Luther lived until 1546, leading the development of Lutheranism. Upon his death, his opponents immediately used their printing presses to spread pamphlets asserting falsely that Luther was a drunk who had died of alcoholism. He had actually died following a stroke. This was one of the earliest known cases of using the media to spread falsehoods, and furthers the analogy between the printing press and the smartphone. The word "propaganda" also links to this era, being derived from the term "Congregation de Propaganda Fide" (Congregation for the Propagation of the Faith), formed by Pope Gregory XV in 1622, to provide training for priests and missionaries

As Europe consolidated into major kingdoms, economic policy became more important as a source of strength and wealth. The commonly held view was that a country was strongest if it produced most of what it consumed, rather than importing it, and that trade is a zero-sum game, in which somebody wins and somebody loses since total wealth was thought to be fixed. Thus, the essential trading objective was to export a higher value of goods than the value of imports, and the difference could be defined by the accumulation of wealth as measured by the precious metals, gold and silver. This excess could then be used to enrich the wealthy and to support their armies and wars of conquest. This economic system, called mercantilism, was prevalent among the emerging European powers from 1500 to 1800.

When Columbus sailed toward the west in 1492, the leading European challengers for power and expansion were in place, and the Roman Catholic Church was still all powerful. The main contenders, although they did not know it as yet, were England, Spain, France, Portugal, and the Netherlands. All faced the Atlantic Ocean although they had no certain knowledge of what treasures its far shores would yield. The technology to navigate safely over long ocean voyages was emerging. All had standing armies well-accustomed to fighting with the best weapons of the day. Their soldiers were immune to the common diseases which would devastate populations not yet encountered. Their royalty aspired to greater wealth and power. They were certain they practiced the one true religion and that God was on their side. They knew their race was superior to all others. Thus, further armed with unacknowledged roots in African DNA; Mesopotamian religion; Chinese gunpowder, printers, and compasses; Hindu number system; Islamic mathematics; and Egyptian paper and ink, Europeans proceeded in the next four hundred years to conquer the unsuspecting and unprepared world.

The Atlantic-facing countries of Europe found it desirable to seek trade routes to their west and south for three fundamental reasons: Trading via the traditional routes through the Middle East became undesirable due to domination by the Muslim Ottoman Empire. Trade was costly, because of the number of middlemen involved and the distances to travel from India and China for spices, silk, and other desirable goods. Finally, European trade was already dominated by the Italian city-states including Genoa, Venice, and Florence.

This led directly to what we now call the "Age of Exploration" or "Age of Discovery," which lasted from the fourteenth through seventeenth centuries. This period could also be aptly called the "Age of Exploitation." However named, the age is defined by its great voyages of discovery, first by Portuguese and Spanish explorers, followed by the English, Dutch, and French. Some of the most significant included the following (there were many others):

1336 – Portuguese expeditions to the Canary Islands

1419–1427 – Portuguese exploration of West African coast

1492 – Columbus reaches the West Indies

1497 – John Cabot explores North America

1498 – Vasco da Gama establishes sea route to India

1513 – Balboa reaches the Pacific Ocean

1519 – Magellan circumnavigates the world

1519 – Cortes conquers the Aztec empire

1522 – Pizarro conquers the Inca empire

1542 – Cabrillo reaches California

1543 – Portuguese travelers are the first Europeans to reach Japan

1579 – Drake explores the Pacific Coast of North America

1606 – Janszoon is the first European to reach Australia

1611 – Henry Hudson searches for Northeast passage to India

Throughout this period, the Europeans took the view that whoever reached an area first could claim ownership for his king, enslave or murder the natives, confiscate whatever wealth they found, and impose their religion on the survivors, all in the name of God, Gold, and Glory. They also felt free to colonize conquered territory and to replace the indigenous peoples with their own. Notable examples include North America, South America and Australia. For a time, Spain and Portugal were the wealthiest countries on Earth, but England and France soon realized that their aspirations required the exploration and conquest of as much of the New World as possible. Although Europe comprises less than ten percent of land on Earth, between 1450 and 1900, Europeans conquered more than eighty percent of the world. Starting from the vantage point of Europe in the aftermath of the collapse of the Western Roman Empire in 400 CE, this would have been impossible to predict.

Europe in the Middle Ages was divided into thousands of competing realms, each subject to attack. Over ten thousand castles were built to provide defense and refuge.

From the fourteenth through sixteenth centuries, European nations were able to explore the world in search of new sea routes to Asia, to conquer and colonize newly discovered lands; to gain wealth, to import spices, and to spread Christianity.

With fewer than 200 men, Spanish Explorer Francisco Pizarro was able to overthrow the Inca Empire in 1532. This was one of many instances of European conquest and colonization of the Americas.

Chapter 15
Modern Times – Revolutions and Ideologies

---✎✎✎---

For he himself has said it,
And it's greatly to his credit –
That he is an Englishman!
W. S. Gilbert and Arthur Sullivan
"H.M.S. Pinafore"
(1878)

It was the best of times, it was the worst of times, it was the
age of wisdom, it was the age of foolishness, it was the epoch
of belief, it was the epoch of incredulity, it was the season of
light, it was the season of Darkness, it was the spring of hope,
it was the winter of despair.
Charles Dickens
(1812 – 1843)

In the history of Homo sapiens, there have been three major epochs. For over 200 thousand years, everyone was a hunter/gatherer involved daily in the search for sustenance and safety. Next, only six or seven thousand years ago, the advent of agriculture converted nearly everyone into farmers. Following in 1760, less than three centuries ago, the third epoch, the Industrial Revolution, began in England, moving most of us off the farm and into the city. The Industrial Revolution, broadly defined, was the transformation from an agrarian and handicraft economy to one dominated by industry and machine manufacturing. In the late 1700's, 90 percent of the population lived on farms; today it is around one percent. If there is to be a fourth epoch, it is altogether unclear what it might be.

The Industrial Revolution not only led to transforming manufacturing, but it altered the vision of how wealth is generated, the rights of individual human beings, and how we should be governed. Modern times can be viewed as a competition among economic systems and systems of government, including all of the struggles and wars induced by them. The competing economic systems have been mercantilism vs. capitalism vs. socialism vs. communism. Monarchies, democracies, republics, social democracies, dictatorships, fascist states, oligarchies, and autocracies are the competing government systems. All of these interact, overlap, and have varying details. This brief discussion of modern times spans the Industrial Revolution, economic systems, forms of government, human rights, the modern structure of nations, and finally an assessment of where the United States stands today.

The Industrial Revolution

The seminal event of the Industrial Revolution was the perfection of the steam engine by James Watt in 1760. His invention was an improvement of earlier versions. The steam engine quickly led to other fundamental applications, including providing power to factories wherever located and powering railroad engines. This helped revolutionize the textile industry, the most important industry in the world at the time. Steam power also revolutionized ocean travel and became the energy source for machines and vehicles of many types. The first railroad using steam locomotives opened in England in 1825, and soon thousands of miles of track spanned England and, thereafter, the world. The demand for track also accelerated development of the steel industry, as well as enabling transportation of freight and passengers.

We can only speculate as to why the Industrial Revolution first occurred in England rather than, for example, China or Islamic Empire, which had been more advanced for most of the previous millennium. Several reasons are postulated: The textile industry in England was producing substantial profits, and wages were high relative to the rest of the world, so that factory owners were motivated to sponsor productivity improvements. The laws in place protected private property and encouraged economic risk-taking. The Renaissance had paved the way for imagination and creativity. There was an emerging class of educated, technology-oriented inventors. Wales had large supplies of readily available coal, but mines had to solve the basic problem of water accumulation, and a solution (i.e., the steam engine pump) was needed. Supplies of iron were abundant, but improved processes increased the demand for coal. Finally, food production increased, requiring fewer farmers, creating a readily available labor pool for factory employment.

Arguably, England, and by extension Great Britain, is the most influential nation in the history of Earth. Isaac Newton is usually considered the greatest scientist. William Shakespeare is most probably the greatest writer. Charles Darwin had what we have called elsewhere the greatest idea – natural selection. Rosalind Franklin, James Watson, and Francis Crick divined the double helix structure of DNA, the basis of all life on Earth. Steven Hawking best understood the nature of black holes. Adam Smith (Scotland) was the father of capitalism. John Locke was the great articulator of human rights. Alan Turing and George Boole defined the concepts underlying digital computing. Peter Higgs conceived the particle named after him (the Higgs Boson) which was an essential contribution to the standard model of particle physics. Even Karl Marx, more infamous than famous, spent most of his life in England.

The Industrial Revolution, which transformed life on Earth, originated in England. For all of its greed and maltreatment of others, the British Empire left indelible imprints in transmitting its language and culture (including cricket, baseball derived from cricket, soccer, rugby, American football derived from rugby, golf and tennis) to much of the world. Had Winston Churchill not found the courage to stand against Hitler, all of Western Europe might still be living a nightmare. These seminal contributions span over 400 years, an unmatched period of invention, creativity, and productivity.

Within a century, the Industrial Revolution spread throughout Europe and Asia. Currently China is the largest industrial manufacturer (29 percent of global manufacturing output), followed in order by the United States (17 percent), Japan (8 percent), Germany (5 percent), India (3 percent), and South Korea (3 percent). The United Kingdom now ranks ninth (2 percent).

In its transformation of civilization, the effects of the Industrial Revolution are pervasive. It has produced a huge migration of people from farms to towns and cities, and transformed their work from farm labor to the many careers necessary in an industrial society. For the first time in history, it has enabled the standard of living of the common man to improve. First the agricultural revolution, then the industrial revolution enabled enormous population growth. It transferred wealth and power in society from land owners to industrialists. Total productivity of goods and services skyrocketed.

The Industrial Revolution also had negative results. Conditions in factories often involved long hours, often 12 to 16 hours per shift, six days a week, at low wage. Children were employed, sometimes as young as five or six years old. There was little consideration of worker health and safety. Women were often preferred as laborers, because they were generally paid about half of a man's wages. Pollution of rivers and the air went unchecked and unregulated. The high demand for cotton, sugar and tobacco from the Americas led directly to the institution of slavery and the involuntary transport of millions of people from Africa.

Economic Systems

At the beginning of the modern era, mercantilism was the predominant economic model with the objective of trade to acquire gold and silver. The Industrial Revolution rendered this economic model obsolete, in favor of what is now called capitalism. The father of modern economics is the Scottish philosopher, Adam Smith (1723 – 1790). He advocated that markets work best when the government leaves them alone. In 1776, he published his book "An Inquiry Into the Nature and Causes of the Wealth of Nations", now regarded as the birth of free-market economics.

In capitalism, a country's industry and trade are controlled by private owners for profit, rather than by the state. The production of goods and services is based on supply and demand rather than through central planning. Under laissez-faire capitalism, individuals are not restrained in terms of where to invest, what to produce, and at what price. A mixed capitalist system includes some degree of government regulation.

Among the beneficial effects of capitalism is the incentivizing of entrepreneurs to increase profit and value in order to maximize their personal wealth. In turn, this stimulates finding the most efficient means of production, such as factory automation, specialized labor, production in an advantageous location, and minimizing labor cost.

A resulting characteristic of capitalism was a society divided into two broad classes; capitalists who defined the rules and kept the profit, and laborers, who worked under conditions set by the capitalist at the lowest possible wage. As an element of the Industrial Revolution, this produced such abuses as child labor, extremely long hours, unsafe working conditions, and subsistence level wages.

In response, the German philosopher Karl Marx (1818–1883) developed the theory that capitalism would inevitably lead to warfare between the capitalist and working classes. In his Communist Manifesto, published in 1848, he promoted that workers should triumph and own the means of production. In communism, all property is publicly owned and each person works and is paid according to their abilities and needs. Communism seeks to create a classless society, by means of revolution, in which there is no private property and totally centralized planning and regulation. The state is responsible to provide work and compensation to every individual. As famously quoted of Marx, "from each according to his ability, to each according to his needs."

Socialism is a variation on the communist theme in which all means of production are owned by the state while still allowing private ownership of property and rewards for personal initiative. Marx is often named as the father of modern socialism as well as the father of communism.

The Russian revolutionary, Vladimir Lenin, adapted Marx's ideas to his own vision, called Leninism, which envisioned a dictatorship of the working class. He successfully led the revolution in Russia in 1917. Mao Tse Tung gained communist control of China in 1959.

The ensuing struggle between capitalism and communism is the defining feature of the late twentieth century, continuing to this day. In fact, communism as envisioned by Marx and Lenin was a failure, largely because communism does not recognize the importance of the profit incentive in inspiring effort and because of opposition by powerful persons, nations of wealth, and other opponents of the communist ideas. Today's Russia and China can no longer be thought of as communist nations.

Although there are no purely communist or socialist nations today, the concept of social democracy has emerged with values similar to a purely socialist economy contained within a capitalist framework. A competitive market economy and private owner-ship of property is supported while providing social programs to benefit the common welfare. Social democracies are sometimes referred to as "welfare states", most common today in much of Western Europe. Even nations which regard themselves as capi-talist states generally have limited forms of social safety net, such as unemployment insurance and national health plans. Today, the main political struggle in many countries, including the United States, is the degree of safety net that should be provided.

Forms and Features of Government

All modern nations are hierarchic in their organization with one or more leaders at the top. In every case the original root of leadership is force, whether resulting from warfare between countries, internal revolution, or terrorism. After the initial leadership is established, successive leaders come to power in a variety of ways, including heredity, election, or power struggle among competitors or survivors.

Most early nations were **monarchies** whose kings' or queens' right to govern was claimed to be divinely ordained and with succession being hereditary. Thus, it was essential that the current royals bear an heir, preferably male, to guarantee orderly succession.

In a **pure democracy,** decisions are made directly by vote of the citizens. Although ancient Greece is described as a pure democracy, power was really shared only by property-owning males to the exclusion of all others, and there were established leaders, notably Pericles. A pure democracy is, of course, impossible in nations with total voters in the millions. Of note is that the word "democracy" does not appear in the United States Constitution.

In a **representative democracy**, most commonly a **republic**, leadership is chosen by direct or indirect vote of eligible citizens, with new leaders elected at specified intervals. Today, 159 of the 193 nations use the word republic in their official names, including North Korea, currently the most undemocratic of nations. The leaders' powers may be limited by one or more representative bodies, such as a senate or parliament, and by a body of law interpreted by an independent judiciary.

In an **autocracy**, there is one person with absolute power to make decisions for his nation. **Autocracy** and **dictatorship** are essentially identical concepts. **Autocracy** and **totalitarianism** are closely related, except that a **totalitarian state** may be governed by a coalition or a single political party rather than strictly by a single individual.

A **kleptocracy** is a government by those whose primary goal is personal gain at the expense of the governed. Contemporary Russia is often characterized in this way. The number of billionaires per trillion dollars of gross domestic product is far greater in Russia than in the United States. Similarly, an **oligarchy or plutocracy** is a country in which a small group, usually of the wealthy, hold most or all political power.

Most leaders encourage **nationalism** and **patriotism,** and citizens are urged to identify with the nation and to support its interests in competition with other nations. Citizens are expected to place national interests above personal interest and to feel superior to other nations. **Xenophobia** and **jingoism** are terms associated with extreme patriotism, especially in support of aggressive intent towards other nations.

Fascism is a political ideology characterized by extreme authoritarianism, private ownership of property, and extreme nationalism. Fascism is enforced by imprisonment or execution of dissidents and minorities, and aggressive intent to expand by conquering or controlling other nations. Germany under Adolf Hitler and Italy under Benito Mussolini are the most notorious examples. A main distinction between Germany under Hitler and Russia under Stalin was that in fascism, private ownership of property is permitted, whereas under autocratic communism, it is not.

Globalism is the concept of finding ways to solve world-wide problems in economics, foreign policy and the environment via international cooperation and mandate, rather than purely nationalistic solutions. For example, no single nation can solve the environmental issues surrounding global warming, pollution, and migration. Many regard the conflict between globalism and nationalism as one of the central issues of the twenty first century.

Human Rights

Our contemporary idea of human rights is that all are entitled to certain inalienable rights simply because we exist as human beings, independent of nationality, sex, race, religion, or any other characteristic. The most basic of these include the rights to life; equal treatment under law; freedom of speech, assembly and religion; and equal opportunity for education and employment.

All civilizations before modern times focused power and wealth among a very small number of people at the top and a very large number of slaves at the bottom, with slaves having no rights to life, liberty or pursuit of happiness. Both Aristotle and George Washington owned slaves who were theirs to do with as they wished. The earliest known example of a major human rights initiative was in Babylon in 539 BCE when Emperor Cyrus the Great freed slaves and declared the people should be free to choose their religion. The concept of rights granted by natural law under God was developed by St. Thomas Aquinas in France in the thirteenth century.

John Locke, 17th century English philosopher, is generally regarded as the father of modern human rights. He wrote that all persons are equal with rights that are God-given and cannot be taken away. The fundamental rights he identified were life, liberty, and property. His work led to the English Bill of Rights (1689), the American Declaration of Independence (1776), and the French Declaration of the Rights of Man (1789).

Today, 194 countries have signed the Universal Declaration of Human Rights (UDHR), issued by the United Nations in 1948, which lists thirty basic human rights. This document has been translated into nearly 400 languages, making it the most translated document in the world. The UDHR attempts to expand human rights to economic rights including "a standard of living adequate for the health and well-being of his family including food, clothing, housing, medical care, and necessary social services."

These rights are not completely achieved in most nations. In addition to the on-going increase in authoritarian governments and their disdain for human rights, important issues are the inherent inequity arising from the uneven distribution of wealth and the assault on Mother Earth.

Virtually all countries today have elections, but in many cases the outcome is preordained by limiting eligible candidates to only one party, intimidation of voters, and restricting voting rights. In the United States today, elections to national office are largely contests to determine who can raise the most money for advertising and who can use social media to propagandize and manipulate well-targeted groups of voters with self-serving and distorted messages.

The wealthiest one percent of the world's eight billion people own almost half of the total world's wealth and the poorest fifty percent of people own one percent of the wealth. Nearly ten percent, about 800 million people live on less than two dollars per day, and about one billion do not have enough to eat on a daily basis. More than three million children die of hunger related causes every year. The other 29 basic human rights are of little value to a starving man, woman, or child. It is estimated that air pollution is a major contributing cause of the deaths of more than ten million people per year, or about one death out of every five worldwide.

In summary, we have come a great distance in recognizing that basic rights accrue to all humans, but there is much yet to be done.

The Development of the Modern System of Nations

It is far beyond the scope of this book to give a detailed account of how every nation developed over the last 500 years. However, it is possible to describe the broad outline of our history in fairly succinct terms. After Columbus reached the New World, the European nations, primarily Britain, Spain, Portugal, and

France, proceeded to colonize the Americas and Australia in the seventeenth and eighteenth centuries. In the eighteenth century, India, and much of Asia and the Middle East were conquered by the British and the French and ruled as part of their empires. The final major area colonized and/or conquered by the Europeans was Africa beginning late in the nineteenth century and continuing until the end of World War I in 1918. Thus, the period of colonial and imperial development spanned a little over four hundred years, from 1492 to 1918.

However, and this should be no surprise, it came to pass that colonists, like those in the Americas, and the conquered, as those in India and China, soon resented their domination from afar, and sought independence. The American Revolution (1776 – 1783) was the first major example, with many more to follow.

In the meantime, the Europeans were engaged in constant conflict among themselves. Great Britain and France alone fought four separate wars in the seventeenth and eighteenth centuries. Napoleon set about to conquer the world for France in the early nineteenth century, ending with his catastrophic invasion of Russia in 1812 and his ultimate defeat by the British at Waterloo in 1815. This ultimately led to the two devastating world wars between 1914 and 1945 which left most of Europe exhausted and impoverished. The result was not only the end of their empires, but also the end of monarchy as a common form of government, replaced by a dynamically changing set of republics and autocracies.

Here, then, is a brief (and incomplete) summary of the major events which yielded the modern world of almost 200 sovereign nations that we live in today:

1775 – 1783   The American Revolution forms the United States.

1789 – 1799   The French Revolution overthrows the French monarchy.

1799 – 1815	The Napoleonic Wars end French imperial aspirations in Europe.
1817 – 1825	Spain's colonies in Central and South America win independence. These include Chile, Colombia, Ecuador, Panama, Venezuela, Peru, and Brazil.
1867	Canada becomes a self-governing entity within the British Empire.
1901	Australia becomes a self-governing member of the British Commonwealth.
1904 – 1905	Japan defeats Russia in the Russo-Japanese war and becomes a world power.
1907	New Zealand becomes effectively independent of Great Britain.
1917 – 1922	The Russian Revolution leads to the end of the Russian monarchy and the founding of the communist Union of Soviet Socialist Republics under Vladimir Lenin.
1919	The European map is redrawn by the Treaty of Versailles at the close of World War I.
1922	Egypt declares independence from Great Britain.
1931	South Africa gains full independence from the United Kingdom.
1932	Kingdom of Saudi Arabia founded.
1933 – 1938	Adolf Hitler comes to power in Germany and initiates World War II.
1937 – 1949	Communists come to power under Mao Tse Tung and form the Republic of China.
1945 – 1946	The United Nations is formed with 51 member nations. (Today there are 193.)
1947	India and Pakistan gain independence from Great Britain.

1948	The State of Israel is founded from British controlled Palestine.
1954 – 1975	Vietnam ends rule by the French and intervention by the United States to form a unified Vietnam.
1988 – 1991	The Soviet Union collapses resulting in 15 new independent nations with eight nations in central Europe also gaining independence.

In summary, today's political world is gathered into 193 distinct and sovereign nations, each with government of widely varying degrees of democracy and authoritarianism. Almost all are governed under written constitutions, of which the American constitution is the oldest. Every economy features its own degrees of capitalism and socialism, with no country either totally laissez faire capitalist nor socialist. Pure communism is a failed system (at least as envisioned by Karl Marx), although some nations (most notably China) are ruled by Communist Parties.

Comprehending the United States in the Modern World

Since I am a proud American and cheer for my Olympic team, it is reasonable for me to attempt to comprehend in an objective way how my country stands in the world. We Americans tend to have a highly positive view of ourselves, justified by our contributions in the last 250 years, and our role as the strongest military power and self-proclaimed defender of democracy and human rights for the past century. Ronald Reagan characterized us in 1980 as "a shining city upon a hill whose beacon light guides freedom loving people everywhere," and nearly every American president since World War II has expressed a similar sentiment. John F. Kennedy made reference to the city upon a hill in a 1961 speech, and the phrase was first used in North America in a sermon given by Puritan minister, John Winthrop, in 1630, quoting Jesus' Sermon on the Mount as given in the Bible (Matthew 5:14).

Many organizations rate all of the world's nations on a wide variety of measures, and we will look at a few of them. In each case, we show the ranking by country from one through five. In addition, we show the rankings of the United States as well as for China and Russia since they are our two principal competitors for power and influence. You can then decide what you think about our standing in the world, and whether there might be a case for a more introspective, modest, and balanced view of ourselves.

In considering this information, be aware that much of it is subjective and open to debate. So, if you find something to question, please do so. The source of all of it is the web, and you can find it easily and measure the credibility of anything you question. Also, if comparing nations on almost any other measure you can imagine is of interest to you, relevant data is probably readily available via a simple Google search.

MEASURES OF PRODUCTIVITY

National Gross Domestic Product (GDP)
1. United States – $20.5 trillion
2. China – $13.4 trillion
3. Japan – $5.0 trillion
4. Germany – $4.0 trillion
5. United Kingdom – $2.8 trillion
(11. Russia- $1.6 trillion)

Manufacturing Output (percent of global manufacturing output)
1. China – 28.4 percent
2. United States – 16.6 percent
3. Japan – 7.2 percent
4. Germany 5.8 percent
5. South Korea – 3.3 percent
(15. Russia – 1.7 percent)

Per Capita GDP

1. Qatar – $129,000
2. Macao – $115,000
3. Luxembourg – $108,000
4. Singapore – $94,000
5. Brunei – $79,000
(13. United States – $60,000)
(57. Russia – $26,000)
(79. China – $17,000)

Number of Billionaires

1. United States – 724
2. China – 698
3. India – 140
4. Germany – 136
5. Russia – 117

Patents Granted (2020)

1. China – 530,127
2. United States – 351,993
3. Japan – 179,382
4. South Korea – 134,768
5. European Patent Office – 133,706
6. Russia – 28,788

MEASURES OF SIZE

Population

1. China – 1.44 billion
2. India – 1.38 billion
3. United States – 331 million
4. Indonesia – 274 million
5. Pakistan – 221 million
(9. Russia – 146 million)

Land Area (million square miles)
1. Russia – 6.3 million
2. China – 3.6 million
3. Canada – 3.5 million
4. United States – 3.5 million
5. Brazil – 3.2 million

MEASURES OF QUALITY OF LIFE

Quality of Life Index
1. Switzerland – 195
2. Denmark – 192
3. Netherlands – 185
4. Finland – 185
5. Australia – 184
(15. United States – 171)
(67. China – 105)
(70. Russia – 103)

Freedom Index
1. Switzerland – 9.1
2. New Zealand – 9.0
3. Denmark – 8.9
4. Estonia – 8.91
5. Ireland – 8.90
(19. United States – 8.73)
(59. China – 8.05)
(126. Russia – 6.23)

Life Expectancy
1. Hong Kong – 85.2 years
2. Japan – 85.03 years
3. Macao – 84.6 years
4. Switzerland – 84.25 years
5. Singapore – 84.07 years
(46. United States – 79.1)
(72. China –77.1)
(113. Russia – 73.0 years)

Average Intelligence Quotient (IQ)

1. Japan – 106.5
2. Taiwan – 106.5
3. Singapore – 105.6
4. China – 104.1
5. South Korea – 102.3
(29. United States – 97.4)
(35. Russia – 96.2)

Number of people imprisoned per 100,000 population

1. United States – 629
2. Rwanda – 580
3. Turkmenistan – 576
4. El Salvador - 564
5. Cuba – 510
(23. Russia – 326)
(128. China – 119)

Percent of population with a college degree

1. Canada – 54 percent
2. Russia – 54 percent
3. Israel – 49 percent
4. Japan – 48 percent
5. Luxembourg – 45 percent
(8. United States – 44 percent)
(42. China – 17 percent)

MEASURES OF MILITARY STRENGTH

Overall strength of armed forces

1. United States
2. China
3. Russia
4. Germany
5. United Kingdom

Annual amount spent on defense
1. United States – $778 billion
2. China – $252 billion
3. India – $73 billion
4. Russia – $62 billion
5. United Kingdom – $59 billion

Number of nuclear warheads
1. Russia – 6255
2. United States – 5550
3. China – 350
4. France – 290
5. United Kingdom – 225

Armed Forces Personnel
1. China – 2.81 million
2. Russia – 1.52 million
3. United States – 1.37 million
4. India – 1.30 million
5. South Korea – 683 thousand

Number of overseas military bases
1. United States – 58
2. United Kingdom – 26
3. Russia – 18
4. France – 11
5. India – 6
(9. China – 1)

Here are a few points of interest about these data.

1. Russia's GDP is surprisingly small, only about eight percent of the USA.

2. China has almost as many billionaires as the United States, and Russia has the highest ratio of billionaires to GDP.

3. To be sure, the United States is a wonderful place to live and we have reason for national pride, but people

in Hong Kong live longer; people in Switzerland have a higher quality of life and more freedom; lots of other countries have higher average incomes and far fewer people in jail; and there are many countries with higher measured average IQ (although the significance of these test results is widely disputed).

4. If I were an American global strategist (which I am not), I would observe that for the first time in our history, America faces a potential adversary (China) which has many more people and a larger standing army, much higher manufacturing capacity, a higher rate of new patent production, and a population not bitterly divided into two irreconcilable camps. To be sure, we have some compensating advantages, and China has many problems of its own, but it would at least pass my mind that we should look for ways to get along, rather than reasons to fight.

Conclusion

In considering the modern world, it is well to remember that it began relatively recently and is a work in progress, changing at breakneck speed. Mankind first sailed around the world almost exactly 600 years ago and flew around it just one century ago. Neil Armstrong stepped on the moon only half a century ago. The internet is just forty years old and the first smartphone is now just a quarter of a century past. Comprehension of modernity comes best in the form of questions rather than answers. Some of the most basic ones are these:

1. Will humanity permanently resist destroying itself with the weapons that it has created through science?

2. Will technology ultimately enable us to preserve our environment or will we render the planet uninhabitable?

3. Will science ultimately provide better understanding of the creation of the universe and the underlying laws of physics?

4. Will the competitive nation state remain the principal form of political organization or will some other form emerge enabling global perspective?

5. What will be the balance of governance between democracy and autocracy?

6. Is there a system of economics to achieve a better balance between capitalism and centralization of wealth, on the one hand, versus social welfare and human rights, on the other?

7. Will technological advances (e.g., artificial intelligence, social media, DNA modification, etc.) improve the human condition or destroy it?

As of now, the only certain answer to these questions is "Nobody knows."

The steam engine, first invented in 1698 in England by Thomas Savery and improved in 1775 by James Watt, was the foundational invention of the Industrial Revolution.

Steam power enabled factories to locate anywhere and provide power to large machines. By 1820, almost all cotton and wool fabric was woven in textile factories.

Steam-powered locomotives enabled railroads to stimulate growth
of the steel industry as well as revolutionize the long-distance transport
of people and commerce.

Steam-powered ships eliminated the dependency of navigation on wind
patterns, greatly reducing travel times, and opening new trade routes.

Chapter 16
Warfare - Bang! You Are Dead

—⊸⧚⧚⊸—

If you are not prepared to use force to defend civilization,
then be prepared to accept barbarism.
Thomas Sowell
(1930 –)

Other animals fight for territory or food; but, uniquely in the
animal kingdom, human beings fight for their "beliefs."
Michael Crichton
(1942 – 2008)

All the principal dimensions of your life, including your existence, were determined by wars and battles fought long ago. If you are an American, you may know that the American republic resulted from the colonists' victory in the Revolutionary War, culminating in the British surrender at Yorktown, Virginia to George Washington on October 19, 1781. You may be less aware that the British Empire owed its existence to the naval victory of the British Fleet over the Spanish Armada in July, 1588, thus defeating the Spanish intent to invade England and overthrow Queen Elizabeth I. The British victory was assisted by favorable winds in the English Channel, which blew many Spanish ships off course. Perhaps if the wind had been blowing differently that day, the Spanish invasion might have succeeded, and the British Empire would not have come to be. If so, North America would have been predominately colonized by the Spanish or the French or both.

French aspirations in the New World were essentially ended in 1803 by Napoleon's need for cash to support European wars. This led to his sale of the Louisiana Territory to the United States under President Thomas Jefferson for the bargain price of fifteen million dollars. Those 828,000 square miles doubled the size of the United States and cleared the way for its expansion across the continent.

If neither of these events had occurred, the principal language of North America would be French or Spanish, the predominant religion would be Catholicism, and the United States would not exist. If your ancestral line traces to the British Isles, you would not exist either.

Warfare has been the constant companion of civilization from its inception, with the first civilization (Sumeria) winning the first recorded war against Elam in 2700 BCE. It has been said that "life is warfare" (Lucius Annaeus Seneca, 4 BCE – 65 CE) and that "history has been the history of warfare" (Godfrey Reggio, 1940 –). If "war" is defined as any conflict that results in the death of more than a thousand people, then a state of war has existed somewhere on Earth more than 90 percent of the time since civilization began. Altogether, there have been hundreds of major wars and thousands of battles and skirmishes which span the five thousand years that civilizations have existed. The United States has been at war for 222 out of its 239 years, or 93 percent of the time, and has participated in over one hundred wars.

In total, about one billion people have been killed by warfare. There are two ways to look at that number. The first, obviously, is that it is a big number and a lot of deaths. On the other hand, one billion is less than one percent of everyone who has lived, so considering the impact of war on the course of civilizations, the relative cost in lives is not very high. More people died from smallpox.

Since the notion of people killing other people seems repugnant, many have wished that humanity could find a way to avoid wars. To date, there has been little to no progress in that regard. When the stakes seem high enough, civilizations are ready to send their sons (and now daughters) to fight to the death. Here are nine common reasons for wars and an example or two of each:

189

Religion – Expansion of Islam (622 – 750), Christian
Crusades to recover the Holy Land (1096 – 1291)
Plundering – Viking Raids (793 – 1066), Mongolian
Empire (1206 – 1368)
Revolution – American Revolution (1765 – 1791),
French Revolution (1789- 1799),
Russian Revolution (1917 – 1923)
Nationalism – Napoleonic Wars (1803 – 1815)
Territorial gain – Mexican – American War (1846 – 1848)
Economic gain – British conquest of India (1858)
Civil war – American Civil War (1861 – 1865)
Revenge – World War II (1939 – 1945)
(i.e., German revenge for WW I outcome)
Defensive – Korean War (1950 – 1953)

It is also true that charismatic figures occasionally arise who are driven to gain their country's domination over all others for the pleasure and power of conquest. Examples include Julius Caesar (Rome, 100 BCE – 44 BCE), Alexander the Great (Macedonia, 356 BCE – 323 BCE), Genghis Khan (Mongol Empire, 1155– 1227), Napoleon Bonaparte (France, 1769 – 1821), and Adolf Hitler (Germany, 1889 – 1945).

Genghis Khan perhaps spoke for this infamous group when he said:

"Man's highest joy is in victory: to conquer one's enemies; to deprive them of their possessions; to make their beloved weep; to ride on their horses; and to embrace their wives and daughters."

In addition to their legacy as conquerors, these five alone were the direct cause of at least 80 million deaths.

Almost without exception, the most dominant civilization in any region is the one with the most powerful military and the willingness to use it. This requires far more than just the most soldiers: There must be wealth that enables equipping and sustaining the warriors. There must be technology to provide weapons at least as effective and preferably superior to those of any adversary. There must be a sense of purpose that provides the population with the will to prevail, and there must also be applause and admiration for the military. This sense of mission might derive from any combination of desire to defend or expand the homeland, to increase wealth and power, to spread an economic or religious ideology, to impose a superior culture on a perceived inferior group, or to capture and enslave a source of labor.

Most successful civilizations convince themselves that they occupy a pre-ordained and divinely-sponsored role in history, and this justifies their domination by force. The earliest Chinese empires concluded that China was the center of the universe and regarded all others as inferior and barbaric. The Romans worshipped their gods of war and believed in their divine assistance. Later Roman emperors declared themselves to be gods. Rudyard Kipling expressed the British Empire's belief that the English bore the "divine burden to reign God's Empire on Earth." European crusaders believed in their divine mission to reconquer the Holy Land. Nazi soldiers wore belt buckles proclaiming "Gott mit uns" ("God with us").

When two armies face each other, the winner usually is the one with the most effective weapons. From that perspective, civilization can be viewed as a continual and unending arms race. Perhaps humankind's first weapon was the closed fist, which is well-designed for striking. By the time of the Sumerians (5000 years ago), weapons of iron and bronze, including maces, swords, spears, javelins, knives and axes were available, and soldiers attempted to protect themselves with body armor, shields, and helmets.

Physical contact of the combatants was required, occurring at the speed that two people could walk or run towards one another. Subsequent weapons development involved increasing the speed

191

of attack, increasing the distance between attacker and defender, and increasing the number who can be simultaneously killed by the fewest attackers. In the time span of only five millennia, we have gone from hand-to-hand combat to potential for millions being killed almost instantly via nuclear guided-missile attack. This seems like a curious measure of progress.

Basic contributions to weapons technology have sometimes come from surprising sources. Leonardo da Vinci, in addition to painting the Mona Lisa, was the inventor of an efficient catapult and designed improved fortifications for castles. Michelangelo regarded himself as being more proficient as a designer of fortifications for Florence than as a sculptor. Albert Einstein's relativity theory led directly to the weapons which hastened the end of World War II.

The evolution of warfare technology occurred in hundreds of incremental steps, but there have only been four fundamental advances which are 1) the use of the horse; 2) the development of explosives and weapons which utilize them; 3) advances in modes of transport including armored vehicles, railroads, tanks, sea-going vessels, and aircraft; and 4) nuclear weapons.

Horses were first used to pull chariots into battle beginning about 1500 BCE. With the development of the saddle and stirrups, warriors could carry weapons into battle on horseback beginning about 900 BCE. Horses were also essential for moving supplies and equipment. The mounted cavalry remained a staple of warfare for over 2500 years ending with World War I (1914 - 1918). These four-hooved warriors were often the victims of combat, with over two million dying in the American Civil War (1861–1865) and over eight million killed in World War I.

The development of explosive weapons began with the invention of gunpowder in China about 1000 BCE. First conceived for fireworks, gunpowder's use in warfare developed slowly. By about 1400, Europeans had learned to use it in cannons and guns. The first modern explosive, nitroglycerin, was invented in Italy in 1846. Dynamite was invented in 1867 by Alfred Nobel in Sweden. Weapons using explosives altered warfare and civilization in fundamental ways. They enabled killing to be accomplished at a distance, so the notion of the heroic warrior weakened. Castles could be obliterated, so they no longer provided refuge from invaders. War became more deadly, so states were compelled to establish and maintain standing armies. Explosive weapons are extremely versatile, including rifles, artillery shells, bombs, land mines, torpedoes, machine guns, hand grenades, depth charges, guided missiles, cruise missiles and others. A favorite tactic of contemporary guerrilla or terrorist fighters is the use of homemade bombs or improvised explosive devices (IED) to disrupt conventional armies. The largest non-nuclear bomb on record weighed about 11 tons and was used by American forces in Afghanistan in 2017.

The industrial revolution saw the development of new modes of transportation, and these are fundamental to modern warfare. By the Civil War in 1861, the United States had a highly developed network of railroads, which were essential to the victory of the North, both in transporting personnel and materiel. The invention of the internal combustion engine led immediately to the development of armored transport vehicles and tanks, ending the role of the horse in warfare. The first powered flight by the Wright brothers in 1903 was followed by the development of the first fighter aircraft used in World War I (1914) and the first jet powered aircraft used by Germany in World War II (1945). Construction of the United States network of interstate highways began in 1956, in part to support military transport and to enable rapid evacuation of cities in the event of nuclear attack.

The advent of nuclear weapons in 1945 introduced a new period in warfare with the potential for total annihilation. Annihilation, that is total destruction of an enemy, is not a new strategy. The Romans destroyed Carthage and the Gauls without compunction, and complete annihilation was Hitler's objective for Jews and Slavs. It is one thing for a leader to imagine totally destroying his enemy. However, it is quite another for him to imagine that in retaliation, his enemy might have the capacity to destroy him, his family, his dog, his country and possibly humanity within a few hours. This new realization led to the doctrine of "mutually assured destruction" (aptly called "MAD") which suggests that the use of nuclear weapons must be avoided because the consequences to everyone are so terrible.

To be sure, no country has employed nuclear weapons since their use by the United States against the Japanese, which ended World War II. The interval from 1945 to the present is one of the longest in history without major warfare directly among leading powers. Many attribute this to the fear of nuclear holocaust. However, at the present time, nine countries have developed a total of about 13,000 nuclear weapons, and so hope for the future depends on no leader ever concluding that mutual destruction is better than defeat, and that no accident will ever occur which unleashes the nuclear nightmare. This is a slender thread, especially projecting ahead a millennium or two.

Even though there have been no major wars since 1945, there have still been almost 300 identifiable armed conflicts around the world, causing over 40 million deaths. This is an area of active scholarly research, with most concluding that the current level of conflict and death is well below historical averages. After all, 40 million deaths represent "only" about half a percent of the total population. The unresolved debate is whether this period presages a continuing downward trend or whether it is just a short-term anomaly.

Two types of warfare which have become more predominant since the end of World War II are guerrilla warfare and terrorism. The term guerrilla is from Spanish and means "little war." Guerrilla wars typically involve small groups of armed dissidents within an existing order who seek revolution by the use of hit and run tactics (ambushes, skirmishes, sabotage and raids) by small groups of irregular forces. The intent is to gradually weaken the existing order and to gain growing support from the populace until the existing order can be overthrown and replaced with one chosen by the guerrilla force. Guerrilla tactics have been known and used since at least the sixth century BCE. The American Revolution (1775 – 1783) was a guerrilla war to a significant extent. Successful guerrilla wars in modern times include Fidel Castro's Cuban Revolution (1956 – 1959), the emergence of Chinese Communism (1946 – 1949), and the Vietnam War (1959 – 1975). In total, there have been at least forty successful guerrilla campaigns since the eighteenth century.

Terrorism involves the employment of indiscriminate acts of violence against noncombatant victims for the purposes of intimidation, inducing fear, and causing permanent psychic damage for political, religious, or ideological goals. Terrorism traces back at least to the sixth century when officials of the Roman Empire were targeted by dissidents. Guy Fawkes' unsuccessful plot to blow up the British Parliament in 1605 was a famous, albeit unsuccessful, terrorist act which is still celebrated annually on November 5 in Great Britain as Guy Fawkes' Night.

By far the largest terrorist act in the history of the United States was the September 11, 2001 destruction of the twin towers of the World Trade Center (WTC) in New York City. Nineteen members of the al-Qaeda network led by Osama Bin Laden hijacked four commercial jets and flew two into the World Trade Center, one into the Pentagon in Washington D. C., with the fourth crashing in Pennsylvania. Almost three thousand people died in this attack, and the day is indelibly marked in the minds of millions of Americans who witnessed it live on television.

The aftereffects of this attack are enormous and on-going. The WTC event initiated almost instantaneous reaction in an attempt to avenge the assault. Domestic anti-terrorist agencies were consolidated into the Department of Homeland Security, which today employs about 120 thousand people. A "Global War on Terror" was initiated which directly led to twenty years of warfare in Iraq and Afghanistan with catastrophic outcome and defeat of the United States. A recent Brown University study estimates that the total cumulative American cost of the war on terror will approach ten trillion dollars after all obligations are met, more than double the cost of World War II in constant year dollars. This constitutes more than one-third of the national debt and equates to over three billion dollars for every life lost in the WTC attack or over thirty thousand dollars for every American man, woman, and child. Internal divisiveness within the United States continues to grow, and more than a million people died in the Middle East as a consequence of the wars there. In retrospect, this was a very costly retaliation.

Since warfare has been such an integral part of the human experience, it has naturally drawn attention from thinkers and philosophers. Three classic works worthy of your attention if you wish further study in this area are "The Art of War" by Sun Tzu, (China, 544 BCE – 496 BCE); "The Prince" by Niccolo Machiavelli (Italy, 1469 – 1527); and "On War" by Carl von Clausewitz (Prussia, 1780 – 1831). Sun Tzu's work is still regarded as a masterpiece of war and business strategy. The term "Machiavellian" refers to political deceit and deviousness. Clausewitz famously concluded that "war is the continuation of politics by other means."

The successful prosecution of war is a highly complex process including many disciplines involving strategy, tactics, and logistics. In addition to the three service academies (West Point, Annapolis, and the Air Force Academy), over forty American universities offer majors in military science, as do other universities around the world.

Military strategy involves the establishment of high-level goals for the outcome of conflict. A good example is the policy of containment established by the United States after the Soviet Union became a nuclear power. This was initially stated in 1947 and called the "Truman Doctrine." The strategic goal was to avoid direct conflict between American and Soviet troops to avert nuclear war, while resisting expansion of Soviet and communist influence. Significant occurrences included the Berlin Airlift of 1948, the Korean War (1950 – 1953), the Cuban missile crisis of 1962 and the Viet Nam War (1955 – 1975). Whether or not this strategy was successful is still a matter of intense debate and unclear long-term outcome, especially considering the rise of China and the continuing antagonism between the USA and Russia. Nevertheless, the Soviet Union dissolved itself in 1991 and is no longer a communist country.

Military tactics refers to the deployment of land, sea, and air forces to achieve near-term military outcomes. Tactics can be categorized as offensive, defensive, and deceptive. An example offensive tactic is "shock and awe" in which massive air attacks are intended to destroy an enemy or city very quickly in order to eliminate their ability or will to fight. This was used by the Germans against Rotterdam in 1940, by the British against Cologne in 1942, and by the Americans against Tokyo in 1945 and Baghdad in 2003. This "shock and awe" tactic was first proposed by Sun Tzu in about 400 BCE. An example defensive tactic is trench warfare, widely used in World War I, as a protective defense against machine gun and artillery fire. In retrospect, trench warfare is seen to be largely ineffective and a contributor to exceptionally high fatality rates among participants. Perhaps the most famous deceptive tactic was the Trojan Horse mythically used by the Greeks to sneak warriors into Troy. A very real deceptive tactic used by England prior to the Normandy invasion of World War II (1944) was the use of fake inflatable tanks and dummy landing craft to deceive the Nazis into believing the invasion would be targeted at Pas de Calais rather than Normandy.

Military logistics is the provision and maintenance of supplies to forces deployed in the field. This can be an enormous and decisive factor, which has been true for all of history. Modern technology enables far larger forces to be deployed over greater distances in less time. In World War II, the largest war in history, more than 80 million men were deployed by the United States, British Empire, Germany, and Japan. They were armed with rifles, grenades, artillery, tanks, personnel carriers, and mines. Over 800 thousand aircraft were built. The United States Navy had over six thousand commissioned ships. There had to be fuel for all of the vehicles. In addition to weapons and transportation, the soldiers all needed food, drink, clothing, and housing every day. They had to have medical attention available. They needed cigarettes, alcohol, and stationery at least occasionally.

The future of warfare is impossible to predict, except that it will doubtless seek to take advantage of whatever science and technology enable. Military applications of technology often occur in advance of their commercial application. Notable examples include jet aircraft, the internet, and satellite surveillance. The first atomic weapon (1945) was applied six years before the first nuclear power plant (1951). The United States military has recently organized specific branches devoted to warfare in space and cyberspace. The potential for weaponization of biological and chemical agents is real and frightening. Hypersonic missiles and aircraft traveling at more than five times the speed of sound are presently indefensible. Drone weapons enable pinpoint attack anywhere without direct pilot risk. Virtually all advanced weapon development is secret and highly classified, so that unless you are a participant with security clearance, it is impossible to know in detail what capabilities actually exist. The three principal adversaries in today's world; the United States, China, and Russia; are active in all of the above areas.

Most historical accounts of warfare focus on the causes, the leaders, the outcomes and the historical significance. However, please take a moment to focus on the perspective of the infantryman just before the battle. Try to imagine yourself as a young male about to charge onto the field of battle with the intent of killing the young men on the other side. Perhaps you are an Athenian about to face the Spartans, a Roman legionnaire about to face the Carthaginians, a Frenchman about to face the English at Waterloo, a German about to go over the top of the trench in World War I, or an American on a landing craft at Normandy in World War II. The circumstances are virtually identical, except for the weapon that your army has given you, whether sword, bow and arrow, musket, or rifle. You might have found yourself in this circumstance because you volunteered out of patriotism, because you were a mercenary being paid, because you were conscripted by your country, or because you were a slave.

Try to imagine your feelings just before you charge onto the battlefield. You know that within the next few minutes you will either kill or be killed. You will either be a triumphant warrior, a survivor, or a corpse. Do you think you would have been exhilarated? Terrified? Purposeful? Many studies have shown that the experience of combat is one of the most profound any human can have and the feelings are lifelong and indelible. Many report exhilaration and pleasure in the act of killing compounded by regret at having done so. They also report that in the last analysis, they were fighting for their comrades and not for their country. On the other hand, one study of World War II soldiers found that about 25 percent of combatants never even fired their weapons.

Historically, almost all human direct participants in combat have been male, and several reasons are proposed to explain this fact. First, males are larger and stronger than females on the average with about 40 percent more upper-body strength and 33 percent more lower body strength. To the extent that combat

Warriors through the ages

Legionnaire, 100 BCE

Minuteman, 1776

Infantryman, 2010

Fighter Pilot, 2020

requires strength, men are better qualified. (However, it should be noted that female participation in warfare is increasing as raw strength becomes less important with modern weapons. For example, there are now female fighter pilots.) Other reasons for male predominance in war have been proposed, if not proven. One is that the male brain matures more slowly than the female brain, so that young males are more likely to undertake risky behavior. Another is that aggressiveness improves a male's chances of attracting female mates, and that aggressiveness overflows to the battle field. Both of these factors would enhance the natural selection of aggressive males. In any event, history proves that armies of young men can always be gathered together to walk away from home to confront other armies of young men with the mutual intent to kill.

Chapter 17
Systems of Belief - Everyone is in the Minority

———— ∞∞∞ ————

Whether one believes in a religion or not, and whether one believes in rebirth or not, there isn't anyone who doesn't appreciate kindness and compassion.
Dalai Lama
(1935 –)

And ye shall know the truth, and the truth shall make you free.
The Bible
(John 8:32)

There are the small questions in life which we face every day. What shall I have for breakfast? What should I wear today? Do I need a haircut? These do not take much contemplation. On the other hand, there are the big questions that face everyone: Who am I? Why am I here? How did I come to exist? Am I going to die and what will happen to me if I do? How am I supposed to behave? When is having sex okay and who can I have as a partner? When is it okay to kill someone? What can I do to influence my fate? Who or what created the world I live in? Is it eternal or not? What should I think and do about people who do not believe what I believe? Who should I look to for guidance? Should I try to convince others to believe as I do? What is true and what is false and what is right and what is wrong? What is good and what is evil?

For most of us, these big questions are answered by our parents and the religion or other system of belief that they passed down. More than three out of five of us adopt one of four major religions: Christianity, Islam, Hinduism, or Buddhism. More than one person out of five do not adhere to any organized religion and are non-sectarian in outlook. The remaining five percent or so are distributed among thousands of other specific systems of belief which have varying degrees of relationship to the first four. For example, Judaism, with Sumerian roots, has been very prominent in history and is the source of both Christianity and Islam, yet Judaism has a membership of less than one percent of world population.

The four major religions arose in two locations. Christianity and Islam have their roots in Mesopotamia in about 1800 BCE, and are called "Abrahamic" after the (possibly mythical) Sumerian prophet who is regarded as the father of monotheism. Although Christianity and Islam have frequently been in conflict, both believe in a single God who created the universe, and they require worship of that God. Similarly, Hinduism and Buddhism both have roots in the Indus River Valley dating from about 2300 BCE and share in common a belief in reincarnation and the spiritual value of meditation. Buddhism developed about a thousand years after Hinduism, and Islam came about six hundred years after Christianity.

Since this book is about comprehension, the purpose of this chapter is to summarize the five major systems of belief so you might better understand those other than your own. Each is presented on its own terms with no judgment or comparison. They all have hundreds of millions of followers. Hopefully, followers can read what is written here and agree their beliefs are represented with a reasonable degree of both accuracy and sympathy. We describe the four religions in historical order followed by a discussion of the non-religious.

Not all members of a given religion believe the same things identically. There are multiple sects of each that have developed different outlooks. What follows is an attempt to describe belief systems without attempting to capture all of the nuances. Each belief is stated as a matter of fact in order to avoid the repetition of "Muslims believe that" and so forth.

HINDUISM (Currently 1.1 billion followers)

Hinduism is the oldest of the four major religions with roots tracing to about 2300 BCE in the Indus Valley, near present day

Pakistan. The word Hindu is derived from the Sanskrit word Sindhu, which is also the root word of the Indus River and of India. The so-called Arabic numerals, which are universally used today, were first developed in India, as were the concepts of the number zero and the decimal point. Hinduism has remained centered in India where almost 95 percent of Hindus reside. It is by far the largest religion in India comprising almost 80 percent of India's total population

Hindus believe the universe had no beginning and it will have no end. Rather, it goes through infinitely many cycles, each trillions of years in duration. There is unity in the concept of Brahman, the supreme God force present in all things. Brahman is in everything and everything is in Brahman. Hindus embrace "atman" or the belief in soul. All living creatures have a soul which is a part of the supreme soul. Every soul experiences many cycles of birth and rebirth with the ultimate goal of reaching the state of moksha, which is the transcendent state attained by being released from the cycle of rebirth. Every soul will ultimately attain a state of moksha and none will be denied. There is no hell since hell would have no God, but God is in everything.

Karma is the sum of a person's actions in this and previous states of existence, and karma will decide each soul's fate in future cycles of existence. Dharma is the religious and moral law governing individual conduct and is an individual's duty of fulfilling the spiritual law for human beings. Dharma is the way of goodness, truth and duty. Dharma is a code of living that stresses good conduct and morality.

Hinduism embraces many religious ideas, and thus it may be seen as a family of religions or a "way of life" as opposed to a single organized religion. A fundamental principle is that a person's thoughts and actions determine his current and future lives. Hindus revere all living creatures and consider the cow a sacred animal. Most, but not all, Hindus are vegetarians. Hinduism advocates happiness and celebrating life and encourages the development of wealth.

One of the early features of Hinduism was the creation of a caste system that divided followers based on their Karma and Dharma. This system had its origins over three thousand years ago, and caste determined every aspect of a person's status in India, ranging from the Brahmin (spiritual leaders) to the Shudras (unskilled laborers) and beneath that the "untouchables." The caste system was banned when India became an independent state, but even today, caste has influence on Indian lives.

Although there is unity in God, there are many individual deities which Hindus choose to worship at their individual discretion. There is no expectation that all Hindus worship the same God or Gods. The most prominent deities include:

Brahma – the god responsible for the creation of the world and all living things

Vishnu – the god that preserves and protects the universe

Shiva – the god that destroys the universe in order to recreate it

Devi – the goddess that fights to restore dharma

Krishna – the god of compassion, tenderness, and love

Lakshmi – the goddess of wealth and purity

Saraswati – the goddess of learning

Hinduism is thus "henotheistic," which means all worship a single deity, Brahman, but at the same time believe in other gods and goddesses. (Brahman and Brahma are distinct entities with Brahman being the Supreme Entity and Brahma is the god of creation.) Followers determine for themselves which gods to worship and believe there are many paths to achieving moksha.

The principal sacred texts of Hinduism are the four Vedas including the Rig Veda, the Samaveda, the Yajurveda, and the Atharvaveda. These were composed about 1500 BCE and contain revelations received by ancient saints and sages. Many Hindus believe their faith is timeless and its basic tenets have always existed. There is no principal prophet, and many Hindus believe Dharma transcends time. The Upanishads, the Bhagavad Gita,

18 Puranas, Ramayana, and Mahabharata are also important Hindu texts.

BUDDHISM (Currently 0.4 billion followers)

The founder of Buddhism was Siddhartha Gautama, who was born to a wealthy Hindu family in East India in 563 BCE. He thus predated Jesus, prophet of Christianity, by over 500 years and Muhammad, the prophet of Islam, by over a millennium. His role in Buddhism is comparable to theirs in Christianity and Islam. In his early years, his family shielded him from all poverty and ugliness, but as an adult, he became aware of the suffering and dissatisfaction experienced by most. In consequence, he developed the belief that the goal of life is to overcome suffering and dissatisfaction, and meditation is the path to awakening and enlightenment. He became known as a Buddha or "enlightened one." He sat for seven weeks under the bodhi tree "feeling the bliss of deliverance." At this time, he awakened to the four noble truths and the eight-fold path leading to Nirvana. He then spent the last forty years of his life teaching and gathering followers prior to his death in 483 BCE.

The heart of the Buddha's teachings is summed up in the Four Noble Truths, which require long contemplation to be fully realized. They are: 1) the truth of suffering – we live in an ongoing state of dissatisfaction. 2) the truth of the origin of suffering – suffering and dissatisfaction arise from causes and conditions. 3) the truth of cessation – there is a possibility of reaching a state where dissatisfaction/suffering has ceased. 4) the truth of the path – there is a path for us to follow that will bring us to such a state, called Nirvana. Achieving the state of Nirvana is the ultimate goal of the Buddhist. Nirvana is the transcendent state in which there is neither suffering nor desire, thereby achieving release from the cycle of death and rebirth.

The path to enlightenment is called the Noble Eightfold Path: 1) Right intention – adopting the path of the religious including non-sensuality, no ill-will and kindness. 2) Right speech – no lying nor rude speech nor causing discord. 3) Right conduct or action – no killing nor injuring, no stealing, no sexual misconduct, and no material desires. 4) Right livelihood – no trading in weapons, living beings, meat, liquor, and poisons. 5) Right effort – preventing the arising of unwholesome states. 6) Right mindfulness – being mindful of the teachings that support the Buddhist path. Awareness of the impermanence of body, feeling, and mind. 7) Right meditation including mindfulness, concentration, and insight. 8) Right view – Our actions have consequences. Death is not the end, and our actions and beliefs have consequences after death including rebirth. Rebirth can be as a human or other animal.

All the teachings of the Buddha are summed up in one word, "Dharma", which is the principle of righteousness and calls on man to be noble, pure, and charitable, not in order to please any deity, but to be true to the highest in himself. Note that the term "buddha" means "enlightened one" and is not exclusive to Siddhartha Gautama, although he is referred to as "the Buddha." Some Buddhists believe that a Buddha is born in each eon of time while others believe that all beings become Buddhas upon achieving Dharma. Once Nirvana is achieved, they will no longer be reborn and will cease to exist.

There is no supreme deity in Buddhism although there is a wide variety of divine beings worshipped in various contexts. The three main Buddhist sacred texts are called Tripitaka which means three baskets. This refers to three collections of the Buddha's teachings including the Vinaya Pitaka – the collection of monastic codes, the Sutra Pitaka – the collection of discourses given by the Buddha and his disciples, and the Abhidharma Pitaka, the collection of higher teachings. There are also many other Buddhist texts, known as "sutras", and the first full printed book is the Diamond Sutra dating to 868 CE. Note that this is 600 years before the Gutenberg Bible.

Following the Buddha's death, his followers organized a religious movement which became the state religion of India in the third century BCE. The religion spread throughout the Indian subcontinent and Southeast Asia. However, Buddhism was nearly extinguished in India by the twelfth century as a result of Muslim invasions. Today, China has the largest population of Buddhists, including about 250 million followers or nearly 20 percent of the Chinese population. There are also large Buddhist contingents in Thailand, Japan, Myanmar, Sri Lanka, Vietnam, Cambodia, South Korea, and the United States.

CHRISTIANITY (Currently 2.4 billion followers)

In Christianity, according to its Holy Bible, there is one omniscient and omnipotent God who created the universe and everything in it. He completed the act of creation in six days, beginning with light on the first day and concluding by creating humans on the sixth day. On the seventh day (the Sabbath) he rested. He created man and woman in his own image in the Garden of Eden, but he expelled them and made them mortal because they were tempted by Satan to commit the original sin of eating the fruit from the tree of good and evil. God told them "by the seat of your brow you will eat your food until you return to the ground; for dust you are and to dust you will return." (Genesis, 3)

Later, God delivered the Israelites from slavery in Egypt, and he chose Moses to give his system of laws for the people, the Ten Commandments. These commandments are: 1) You shall have no other gods before me; 2) You shall make no idols; 3) You shall not take the name of the Lord thy God in vain; 4) Keep the Sabbath Day holy; 5) Honor your father and mother; 6) You shall not murder; 7) You shall not commit adultery; 8) You shall not steal; 9) You shall not bear false witness against your neighbor; 10) You shall not covet.

God created heaven and hell and judges every person to

determine who will enjoy eternal bliss in heaven and who will endure eternal damnation in hell, in accordance with their obeying his laws and repenting their sins. "Depart from me, ye cursed, into everlasting fire, prepared for the devil and his angels." (Matthew, 13:42).

Later, in about 5 BCE, God sent his only Son, Jesus Christ, to be born of the virgin Mary in fulfillment of the messianic prophecies in the Old Testament. As the Messiah, Jesus was anointed by God to be the savior of humanity. During his brief lifespan of 36 years, Jesus was baptized, taught the gospel (i.e., "good news"), and performed many miracles, before being crucified for blasphemy after trial by the Roman governor, Pontius Pilate, at the behest of the Jewish elders, most probably on Friday, April 3, 33 CE. Jesus was resurrected from the dead three days later, on the first Easter Sunday, proving that death is not final and life is eternal.

The core of the Christian religion is that by belief in the death and resurrection of Jesus, all sin can be forgiven and salvation granted together with eternal life in heaven. "If you declare with your mouth, 'Jesus is Lord, and you believe in your heart that God raised him from the dead, you will be saved." (Romans 10:9-10)

The sacred scripture of Christianity is the Holy Bible, which tells the history Earth from its creation through the early spread of Christianity. The Bible includes the Old Testament and the New Testament. The Old Testament comprises thirty-nine books corresponding approximately to the Hebrew Bible which was written primarily in Hebrew between about 1200 BCE and 100 BCE. The New Testament was originally written in Greek and records the teachings of Jesus and his followers. It includes four Gospels, the Acts of the Apostles, twenty-one epistles, and the Book of Revelation. Some Christians regard every word as being literally true, while others believe that it must be interpreted. The Gutenberg Bible was the first book ever printed in Europe, and the Bible has been written and rewritten many times. It is the most sold book in history with about five billion having been printed. About 100 million additional copies are printed every year.

In the years following the crucifixion, Christ's apostles developed a following which predominately remained a small and persecuted sect until the Roman emperor, Constantine, converted to Christianity in 337 CE and his successor, Theodosius, made Christianity the state religion in 380 CE. This enabled Christianity ultimately to become the predominant religion throughout Europe and via colonization to the Americas. The spread of Christianity has been aided by its on-going spirit of evangelism. At the present time, there are more than 300 thousand missionaries of various Christian denominations all over the world.

Of the total of 2.4 billion Christians, 37 percent are in the Americas, 26 percent in Europe, 24 percent in sub-Saharan Africa, and 13 percent in Asia-Pacific. This total is forecast to continue to grow with a projected total of 2.9 billion in 2050.

ISLAM (Currently 1.9 billion followers)

Islam is the complete and universal faith that was partially revealed many times by prophets including Adam, Abraham, Moses and Jesus. The Quran, as revealed to the prophet Muhammad, is the final revelation of God.

The "five pillars of Islam" are the fundamental practices that are obligatory acts of worship for all Muslims. First is the assertion of faith. "There is no god but God and Muhammad is the messenger of God." This statement is normally uttered in Arabic. Second, a Muslim must pray five times a day at set times with the individual facing Mecca. Each prayer has a series of set positions and "Allah is great" is said after each prayer. Prayers are performed at dawn, noon, afternoon, evening and night. The third pillar is giving or charity, usually about 2.5 percent of an individual's income. Fourth is fasting during the holy month of

Ramadan. Fasting is required from dawn to sunset during which food, drink, and sex are prohibited. The fifth pillar is that once in his lifetime, every Muslim must make a pilgrimage to Mecca during the twelfth month of the lunar calendar.

Each Muslim is responsible to God for his own actions and salvation. Nothing will save him from his own sins. Any person can become a Muslim by stating his belief in the singularity of God and in Muhammad as His prophet. There will be a final judgment with the righteous rewarded in paradise and the unrighteous punished in hell. Circumcision of males is required in Islam to represent submission to God's will and the submission of base passions to the spiritual requirements of Islam. Both alcohol and pork are forbidden.

The sacred scripture of Islam is the Quran, which consists of 114 chapters and is the word of God as given to the archangel Gabriel who revealed the word to the Prophet Muhammad. To Muslims, the Quran is infallible. The period of revelation from Gabriel to Muhammad began in 610 and continued for the balance of Muhammad's life until 633. The Quran uses the term "Allah" for God emphasizing Allah's supreme divinity. Since Muhammad was illiterate, he recited the words of the Quran to his followers who recorded them. Many Muslims today memorize part or even all of the Quran.

The old and new testaments of Judaism and Christianity were written by prophets, but Muhammad was the last and greatest prophet, and so the Quran supersedes all the rest. The Quran is God's final revelation to humanity before the end of the world comes. Everything Muhammad said or did came from divine inspiration. His words and acts are designated "Hadiths" and are second in importance only to the Quran. Together, the Quran and Hadiths make up the sharia, which means holy law. In nations ruled by Islam, the sharia constitutes legal authority and judges are tasked to interpret it.

The prophet Muhammad was born in Mecca in 570 and

was orphaned shortly after his birth. He began receiving the revelations of the Quran from the angel Gabriel at the age of forty, and this continued for the balance of his life. While still in Mecca, he preached his teachings emphasizing the primacy of a single God. For this, he was persecuted by the rulers of Mecca, and as a result he and his followers emigrated to Medina in 622 where he was able to establish both political and religious authority. In 629, he and his followers reconquered Mecca, and by the time of his death in 632, he gained control of all Arabian tribes in a single religious and political unit.

This union of religious and political power served Islam well following Muhammad's death, and the Islamic empire quickly spread from Spain in the west to the borders of China in the east. By 750, it was the largest empire the world had yet seen (including the Roman Empire). Europeans were seen as unwashed and primitive barbarians. After the death of Muhammad, Islamic leaders were called caliphs, and Baghdad was established as the hub of the caliphate. The period from the seventh to the thirteenth century is called the Golden Age of Islam. During this period, Baghdad was the center of trade and culture, similar to the role Athens had played a millennium earlier. In fact, their translation of the works of the Athenians preserved them for posterity. Otherwise, the works of Aristotle, Plato, Socrates, et.al. would have been lost forever. Fundamental advances in mathematics, science, and medicine were achieved. The words algorithm and algebra are derived from Arabic. The use of paper was mastered and the largest library in the world was found in the Grand Library of Baghdad, also called the "House of Wisdom." This golden age lasted until 1258 when Baghdad was destroyed by invading Mongols.

Islam is currently the predominant religion in North Africa, the Middle East, Indonesia and Malaysia. This total is expected to continue to grow to 2.8 billion In 2050, making it the fastest growing religion. Some forecasts predict that Islam will surpass Christianity by 2070.

NO RELIGION (Currently 1.2 billion)

Taken in total, the number of non-religious people on the planet is about the same as the number of Hindus. Like all of the others, the non-religious come in a variety of categories including humanists, agnostics, atheists, skeptics, and others. For this discussion, all are converged into the humanist name, because all have the characteristic of not practicing any organized religion. Agnostics do not dismiss the possibility of the existence of any god, but believe that there is little or no empirical evidence supporting that belief. Atheists hold the more dogmatic belief that there definitely is no God.

Humanists believe they are each responsible for themselves, and this life on Earth is the only life they can count on. Reason, logic, and evidence are the valid bases for knowledge. There are values, ethics, and a sense of right and wrong which do not depend on judgment, punishment or reward from any supernatural force. In fact, there is no sufficient evidence that the supernatural exists. There are no miracles.

Humans are responsible for choosing the meaning of their own lives, and this can take many forms ranging from commitment to family or nation, service to others, personal accomplishment, advocating for an organization, or working towards some public or private goal. Moreover, the option to accomplish a meaningful life is open to everyone. Bertrand Russell defined the good life as one "inspired by love, guided by knowledge" and A. C. Grayling said that a good life is "the life that feels meaningful and fulfilling to the one living it."

Both freedom of religion and freedom from religion are essential. You may believe whatever you like so long as you harm no one, but you must not seek to alter other lives based on your beliefs that are not shared.

Among the beliefs shared by most humanists are these: 1)

the inherent worth and dignity of all people; 2) the value of science and the scientific method; 3) individual responsibility for choices and behavior; 4) the preservation of nature of which humans are a part; 5) the reality of the evolution of species and natural selection; 6) service to others and social justice; and 7) celebration of human achievement in the arts, literature, and science.

The Greek philosopher, Protagoras, was an early proponent of humanism when he said, in 440 BCE, "About the gods I am able to know neither that they exist nor that they do not exist nor of what kind they are in form for many things prevent me for knowing this, its obscurity and the brevity of man's life." Socrates was an early martyr of humanism when he was executed in 399 BCE for heresy for "failing to acknowledge the gods that the city acknowledges." Later a strong humanist movement arose during the Renaissance in the sixteenth century, emphasizing the importance of the arts in human life. For many leading philosophers, reason and science replaced God as the means of understanding the world. Charles Darwin's theory of natural selection in the nineteenth century gave momentum to humanism by explaining the evolution of species of which Homo sapiens is only one. In our time, many leading scientists including Albert Einstein, Stephen Hawking, Carl Sagan, Neil deGrasse Tyson, E. O. Wilson and many others are outspoken humanists. A recent poll found that over 90 percent of the members of the National Academy of Sciences do not believe in a personal God.

Altogether, the non-religious population in the world amounts to about 16 percent of the total with concentrations of more than 25 percent in China, the United States, most European countries, Japan, Australia, and others.

Wrap-up

These five systems offer three basic bottom-line choices. The Abrahamic religions offer eternal life, but with the cautionary note that if you fail to gain's God's grace, then you will live in permanent torment rather than bliss. The Indus River religions offer you a long cycle of births and rebirths terminated by Nirvana or Enlightenment. Humanism offers the most empirical evidence via the scientific method, but offers no certainty in this life and no afterlife.

The prevalence of systems of belief change over time, and we have taken a snapshot in the year 2021. However, systems based on animism were prevalent for the first two hundred thousand years of human experience, followed by the predominance of polytheistic systems from the Sumerians to the Romans which lasted over five thousand years, followed by the development of the Abrahamic and Indus river religions which to date have prevailed about two thousand years. Humanism as a system of belief was not possible until the emergence of science, and thus is only 600 years old. The next 250 thousand years will pass as quickly as the last 250 thousand, and no one can know the future path of systems of belief for certain.

The purpose of this chapter has been to provide summary descriptions of the world's leading systems of belief in order to better comprehend the ones we do not embrace. There is the strong tendency to reject the ones we do not practice out of hand on the grounds that what we believe is right and what everyone else believes is wrong. You might want to be cautious in your certainty, however, since every one of these systems is supported by hundreds of millions of people, and we are all in the minority.

Chapter 18
The Biosphere - To Be or Not to Be

———∞∞∞———

What mankind must know is that human beings cannot live
without Mother Earth, but the planet can exist without humans.
Evo Morales
(1969 –)

The outlook wasn't brilliant for the Mudville nine that day:
The score stood four to two, with but one inning left to play.
And then when Cooney died at first, and Barrows did the same,
A pall-like silence fell on the patrons of the game.
Edward Lawrence Thayer
(1863 – 1940)
(From "Casey at the Bat")

We are all children of Mother Earth. Over the centuries, we have increased poor Mom's workload a lot, as she has to support so many more of us, and we are so much more demanding and messier than we once were. By the end of the hunter/gatherer era there were about 25 million of us, and now there are almost eight billion. Every second, Mother Earth now witnesses the birth of four human babies and attends two funerals. We used to live about 30 years, but now we live close to 80 years. Currently, the average person consumes three kilowatt hours of power and generates almost two pounds of trash per day. We produce nearly a trillion pounds of trash every two months! Mom's workload from humans has increased by a factor of a million or more in the last 6000 years, and she is feeling the stress. She needs our help, or else she is going to have to be rid of us.

All life on Earth resides in a tiny volume of space which is called the Earth's biosphere. The biosphere has three major components: the geosphere (i.e., the solid surface); the hydrosphere (i.e., all the water on Earth's surface); and the atmosphere (i.e., the envelope of gases surrounding Earth). For humans, the habitable volume includes less than 30 percent of Earth's surface area and the thin layer of oxygenated atmosphere, which is only about five miles thick. The biosphere amounts to about one part in a thousand of the volume of Earth.

The question of whether life can alter the biosphere was resolved over three billion years ago, when photosynthesis by plant life created the oxygenated atmosphere which enabled life. Therefore, there is no question that a single species (Homo sapiens) could alter the biosphere in such a manner as to make the planet less habitable or even uninhabitable. All three components of the biosphere are essential to humans, but over virtually all of our history, we have been ignorant of our own capability to impact our environment. With the advent of the Industrial Revolution and the rapid growth in human population, we began to massively pollute all three elements of our world to such a degree that our own long-term survival is now in question. Our air is warming, our forests are afire, and our rivers and oceans are overwhelmed with human waste. By polluting the environment and otherwise destroying animal habitat, we have initiated a sixth extinction event, called the Anthropocene Era, with the potential to eradicate most living species, including our own.

Most attention has focused on global warming arising from the pollution of the atmosphere. The average temperature at the surface of Earth is determined by the balance between heat radiated to Earth from the Sun and heat reflected from Earth into space from the atmosphere. This effect of the so-called "greenhouse gases" is to trap heat in the atmosphere and reduce radiation to space, thereby causing the temperature of the atmosphere to increase. The major greenhouse gases that result from human activity are carbon dioxide, methane, and nitrous oxide. Carbon dioxide arises primarily from the burning of fossil fuels (coal, oil and gas). The principal causes of methane in the atmosphere are leaks from fossil fuel extraction and mining, the stomachs of cattle and other agriculture activities, and landfills. Nitrous oxide in the atmosphere derives primarily from large-scale farming with nitrogen-based fertilizers and cattle ranching.

Carbon dioxide in the atmosphere has increased from 280 parts per million (ppm) just before the Industrial Revolution to 419 ppm in 2021, the highest level in human history. The concentration of methane has more than doubled since preindustrial times, and the global concentration of nitrous oxide has increased by 30 percent in just the last 40 years. There is no question that these changes have all resulted from human activity and that the average surface temperature has increased by about 1.0 degrees Centigrade (1.8 degrees Fahrenheit) since the late 1800's. Moreover, the rate of change is increasing rapidly, with the average yearly increase since 1980 more than double the average in the century before.

Left unchecked, the effects of uncontrolled temperature increase will be cataclysmic to life. There have been many analyses and forecasts which differ in detail. However, the consequences of a temperature rise of as little as four degrees Centigrade would include the flooding of coastal cities affecting hundreds of millions of people, decreased food production as farm land becomes desert, heat waves rendering areas near the equator uninhabitable, widespread water scarcity, increased frequency of high intensity storms, and acceleration of the on-going extinction event.

In the near future one or more "tipping points" could be reached which would cause even greater change. For example, if the ice sheets of Antarctica and Greenland were to collapse and melt, worldwide sea levels would rise more than 30 feet. Then, as the amount of white surface area covered by ice decreases to be replaced by darker land surface, the reflection of heat from the atmosphere would decrease, further exacerbating temperature increase. Although it would take about 5,000 years for all five million cubic miles of ice on Earth to melt, the average

218

temperature could increase by 40 degrees Centigrade, rendering the planet uninhabitable. Although that may seem like a long time, it is less than the interval from the Sumerians to today.

Notwithstanding that most official attention has been given to climate change arising from atmospheric pollution, humanity's potential impact on the hydrosphere (oceans, lakes, rivers, and groundwater) is as dramatic. Over 97 percent of Earth's water is salt water in the oceans; two percent is stored as fresh water in glaciers and ice caps; and one percent is fresh water in lakes and rivers. The main sources of water pollution include climate change, resulting from the excess carbon dioxide in the atmosphere, industrial and mining waste, agriculture, leaks and spills of fossil fuels, sewage, and wastewater.

As the temperature of the atmosphere increases, so does the temperature of the water, and about a quarter of the excess carbon dioxide is absorbed by the ocean. This makes the ocean more acidic, causing it to be more difficult for many marine organisms to form their shells and skeletons. Perhaps most notably, coral reefs around the world are disappearing, with one forecast that they will be totally gone by 2050 as will many other marine species.

Modern industry produces almost eight billion tons of waste per year, and much of it goes directly into rivers and oceans. Eight billion tons equals one ton each for every human on Earth. Plastic waste in the oceans has emerged as a notable problem. An estimated eight million tons of plastic enters the oceans every year, and plastic takes hundreds of years to decompose. There are now over five trillion pieces of plastic killing over a million marine animals per year either from ingestion or entrapment. By 2050, the total weight of plastic in the oceans may exceed the weight of all of the fish. As the plastic follows ocean currents, some has accumulated into at least five great, tightly-packed, "garbage patches." One of these, the "Great Pacific Garbage Patch" is located halfway between Hawaii and California and covers an area twice the size of Texas.

Agriculture is also a major cause of polluted oceans and rivers. Plowing large tracts of land results in runoff of soil, containing fertilizers, pesticides, and all waste associated with animal agriculture. The scale of animal waste creates large volumes of nitrogen together with the antibiotics and hormones which are used to stimulate rapid animal growth. The effect of nitrogen in the oceans stimulates the growth of "algae blooms" (i.e., uncontrolled growth of algae) which create dead zones (called "zones of hypoxia") where there is no oxygen to support marine life. One study found over 400 zones of hypoxia around the world ranging in size up to 63 thousand square miles. Algae blooms can also produce toxins that can kill fish, mammals, and birds.

Oil is another major source of pollution of the oceans including occasional major accidents with shipping or drilling. An estimated 100 million tons of oil reaches the marine environment each year. Surprisingly, major accidents account for less than ten percent of the oil in the oceans. The biggest source is natural seeps of oil beneath Earth's surface which accounts for about half of the total. The next largest source is operational discharges from ships and from land-based vehicles and industrial processes. Among the many negative effects that oil has on marine life are loss of insulation capability of fur and feathers, inhalation, which can affect lungs and reproduction, and poisoning. In total, hundreds of species and millions of marine animals have died from oil pollution.

Another major source of water pollution is that over 80 percent of the world's sewage is discharged untreated. Many components of human sewage are harmful including bacteria, viruses, pharmaceuticals, heavy metals, pesticides, nitrogen and hundreds of other chemicals. These discharges are not only harmful to marine life. About four million people die every year from diseases caused by polluted water. At least two billion people around the world do not have access to clean water.

220

Finally, we come to the geosphere, which is the element of the biosphere where we build our homes and raise our children. At the time of the first agrarian civilizations, about two-thirds of the arable surface of the Earth was occupied by forests, and about four percent was consumed by farms. Dense forests bring many important benefits to the biosphere. First, they generate oxygen for us to breathe. One mature tree generates enough oxygen to support about five people. Trees also absorb carbon dioxide which today is a principal cause of global warming. They also remove other air polluters including carbon monoxide and sulfur dioxide. Trees help generate rainfall by capturing groundwater and releasing it into the atmosphere. Large forests can influence rain patterns for thousands of miles. Forests provide habitat for thousands of species that provide many chemicals used in modern medicines.

However, the need to feed eight billion humans has been addressed primarily by converting forests into farms. Today over half of the forests have disappeared and the percentage of arable land dedicated to farming has increased from four percent to about 38 percent. Of the six million square miles of forest that once existed worldwide, only about 2.5 million square miles remain. Presently, deforestation continues apace with about fifty thousand square miles destroyed each year. Although the rate of deforestation has decreased slightly in the last decade as concerns have grown, at the present rate virtually all of the viable rain forests will vanish by the end of this century.

The consequences of loss of rain forests are of the same magnitude as climate change. Deforestation is often accomplished by burning off the land, which adds greatly to the release of carbon dioxide. This presently accounts for about 10 percent of all carbon dioxide emissions, more than all automobiles and trucks. The 80 percent of land animals and plants living in forests are threatened with extinction. Weather patterns are influenced, and deforestation amplifies soil erosion, dust storms, and flooding.

The need to continuously increase farm land is driven by the need to feed livestock, which consume three-quarters of agricultural land while producing only 17 percent of human caloric consumption. In the last half-century, the number of people has doubled, but the consumption of meat has tripled. Each year we raise and slaughter 50 billion chickens, 1.5 billion pigs, 500 million sheep, 400 million goats, and 250 million cattle to feed the three-quarters of the world population who are carnivores.

The consumption of meat probably played an essential role in our evolution. Hominids first began to scavenge meat over two million years ago, and the high-energy protein of meat may have contributed to our brains becoming large and complex. Changes in our digestive tract initiated by meat consumption may have made more energy available for our brains.

The disadvantages of the production and consumption of meat in the modern world may indicate that we need to change our meat-eating habits. Far less land would be required for agriculture enabling forests to be preserved. The large amounts of fresh water required both for the animals directly and for the food that feeds them would be conserved. About twenty-five calories of feed is required to produce one calorie of beef. Additionally, animals themselves are large contributors to pollution of the environment, accounting by themselves for about 15 percent of greenhouse gas emissions. Animal manure generates large quantities of methane and contributes greatly to the contamination of rivers. Finally, current studies show that a vegetarian diet is healthier, and vegetarians have a lifespan several years longer than those with a meat-centered diet.

For many of us, including me, the degradation of the biosphere has become a personal experience instead of a theoretical one. As I write this, half a million people in Louisiana have been without power for many days in the aftermath of a hurricane. I live in a wooded area of Northern California which is experiencing an unprecedented rate of wildfires as a result of prolonged drought and high temperature. A recent blaze came within three miles of my home, and it was a matter of chance that the wind blew the front in the other direction. The air quality index reached a level above 350 on a scale where 150 is deemed unhealthy for all ages. Our power company finds it necessary to occasionally shut down power for days at a time to prevent the power grid from causing more fires, and when that happens the temperature inside my normally air-conditioned home can exceed ninety degrees Fahrenheit. I already have invested $15,000 to generate my own solar power and may need to spend another $5,000 to provide backup power. Because of the continuing drought, a nearby community is having to purchase water by the truckload just to provide drinking water for its citizens.

Perhaps the gravest question ever presented to Homo sapiens as a whole is whether we can preserve the habitability of Earth in the face of our own profligacy. There is not much time to change our behaviors, certainly less than a century and perhaps as little as a decade. If we do not, then whether you believe in the Armageddon of an omnipotent God or the semi-omniscience of modern science, the end times are upon us, whether a century from now or a millennium.

Our evolution and our civilization have not prepared us very well for this task. Most of us are focused on our own day-to-day life without much knowledge of what lies ahead; we are citizens of sovereign nations, self-serving and competitive; and we are unable or unwilling to acknowledge our impact many thousands of years from now.

One problem in gaining support for biosphere-friendly policies is that at the individual level it is difficult to recognize the impact of our own behavior. If I run my air conditioner an extra hour or throw my battery in the landfill or burn an extra gallon of gasoline in my car, it is hard to imagine that it matters much. Unfortunately, if you multiply those by eight billion, the differences are huge. None of us wants to be told that we are part of the problem and that we must change our habits and lifestyle. For many, it is easier and more comforting to listen to anyone who assures us that climate change is a hoax, and everything will be all right.

A second problem is that it is difficult to support long term policies which cause short term pain. If you and your family have earned your living for three generations as coal miners, it is not congenial to be told that you should find another career in another place. Even worse, if you are chairman of the board of an oil company, you have hundreds of millions of dollars to invest in justifying your activity via lobbying and advertising since your multi-million-dollar bonus and your continued employment depend on increasing next quarter's profit by greater oil sales. Even worse, if you are Prime Minister of an oil producing nation, your re-election and your power may depend on growing the international market to avoid reducing wealth and increasing unemployment, and to support your national interests and your army.

Yet a third fundamental problem is that it is difficult to justify large expenditures today in the hope of uncertain long-term benefit or for the benefit of others far away in space or time. How much would you be willing to increase your taxes on the chance of enabling a better life for people five hundred years from now? I suspect that for most people the answer is "not very much."

Globalization and capitalism have produced a world-wide economy which is predicated on stimulating unending growth. Every nation is focused on what it can do to improve its own standard of living by the ever-increasing use of energy, the increased consumption of natural resources, and the accompanying accumulation of waste. The concentration

The concentration of carbon dioxide and other greenhouse gases reached record highs in 2020. Average global temperature has increased significantly since the beginning of the industrial revolution as evidenced by loss of polar and glacier ice, lengthening of wildfire seasons, loss of coral reefs, and others.

Eighty percent of the world's wastewater is dumped untreated back into the environment polluting rivers, lakes, and oceans. More people die annually from unsafe water than from all forms of violence.

Rain forests are crucial in stabilizing climates, removing carbon dioxide from the atmosphere, and providing homes for more than half of terrestrial animal species. Rain forests are being cleared at the rate of 30 thousand square miles per year and, at that rate, may disappear within a century.

As their habitat is destroyed, more and more species may become extinct. The current rate of extinction of species is comparable to the previous five mass extinction events.

More than 16 thousand species are presently identified as "endangered," that is they face extinction in the near future. Among the endangered are orangutans, tigers, Asian elephants, and blue whales.

Remaining population of orangutans is about 50 thousand.

Remaining population of tigers is about four thousand.

Remaining population of Asian elephants is about 30 thousand.

Remaining population of blue whales is less than 25 thousand.

of wealth in a few countries leaves billions of people in other countries with a low standard of living. At the same time, the few ultra-rich individuals aspire to be even richer.

If there is a ray of hope, it is the growing number of organizations and movements that recognize the danger and are attempting to take action to preserve the planet. The Environmental Defense Fund, the Nature Conservancy, the Natural Resources Defense Council, the Trust for Public Land, and the Sierra Club Foundation are among the leaders. There are hundreds more. You might consider finding one you can support.

In international politics, there is recognition of the potential for permanent climate change as a result of greenhouse gases. The Paris Climate Accords is an international treaty signed in 2015 by 195 nations setting the goal of reducing carbon dioxide emissions to hold atmospheric temperature increase relative to preindustrial levels to 1.5 to 2.0 degrees centigrade. Although this is a positive first step, it still embodies fundamental limitations in that each nation sets and monitors its own goals; there is no mechanism for enforcement; no penalty for failing to meet goals; and no restriction from recalcitrant nations withdrawing altogether, as the United States did in 2016, albeit rejoining in 2021. Also, the agreement does nothing to address the pollution of the geosphere or hydrosphere nor of other greenhouse gases. In the intervening years from 2016 to 2020, there were in fact modest reductions worldwide in carbon dioxide emissions, but far less than needed to meet the goal of limiting temperature increase to 2 degrees Centigrade.

At the outset of this chapter, I quoted the beginning of a famous poem written in 1888 by Ernest Thayer about a legendary baseball batsman, Casey, who came to the plate in the final inning with a chance to win the game with a home run. Alas, the ending is not a happy one:

"Oh, somewhere in this favored land the sun is shining bright; the band is playing somewhere, and somewhere hearts are light, and somewhere men are laughing, and somewhere children shout; but there is no joy in Mudville – mighty Casey has struck out."

Our species may be in our ninth inning, and like the Mudville nine, our outlook is not brilliant. And like Casey, we may be down to our last out before the game ends. But who knows? We can hope for a home run as if our great-grandchildren's lives depend on it, which in fact they do. Humanity may somehow hit a home run, and then men can still laugh and children shout a century and a millennium from now.

Part 4
From Comprehension to Conjecture and Conclusion

Chapter 19
The Meaning of Life – It Is Your Choice

———— ✺ ————

In the first three parts, we attempted to comprehend the universe and our world over the 13.7 billion years since the Big Bang, based on the results of science and the events in human civilization. Now in Part 4, we focus on three topics that involve conjecture and judgment: the meaning of life; the future; and finally, the main ideas that define our civilization and how they were generated.

Discerning the meaning of life has been a central preoccupation of Homo sapiens from our origin. Early human societies independently developed animist systems of belief; establishing meaning is a major role of all modern religions; and it is usually regarded as the central issue of philosophy. Not surprisingly, there is no universally accepted concept of the meaning of life, and this chapter explores various alternatives.

Using their own words, advocates of the main alternatives articulate their theses accompanied by my occasional comments. Then, I offer my own definition, which is derived from my grasp of all of the others. In the last analysis, it is up to you to assess the meaning of your life, and I suggest a way for you to do so. Perhaps you already have. You can either sign on to one of those offered, or you can form your own. It is also fair not to think about it, and just live your life as it occurs.

We start with the most famous of the philosophers of ancient Greece, who are still regarded as some of the greatest thinkers in all of history.

Happiness is the meaning and the purpose of life, the whole aim and end of human existence.
Aristotle
(384 BCE – 322 BCE)

The unexamined life is not worth living.
Plato
(428 BCE – 347 BCE)

For most people in the world, religion goes far toward defining the meaning of their life.

For Christians and Muslims, it is to have unquestioned faith in one God and His prophet, either Jesus or Muhammad, and to obey His commandments.

For God so loved the world, that He gave his only begotten Son, that whosoever believeth in Him should not perish, but have everlasting life.
The Bible, John 3:16

We have sent a messenger to every nation saying "Worship God and avoid false gods."
Quran, 16:36

Hindus and Buddhists believe in a never-ending cycle of birth and rebirth and seek a state of enlightenment through contemplation and good works.

For the soul, there is never birth nor death.
Nor, having once been, does he ever cease to be.
He is unborn, eternal, ever-existing, undying and primeval.
He is not slain when the body is slain.
From the Bhagavad Gita (author unknown)
(ca. 300 BCE)

Suffering is temporary.
Enlightenment is forever.
The Buddha
(Siddhartha Gautama)
(ca. 563 BCE – ca. 483BCE)

At the other end of the spectrum of the meaning of life are those who have concluded that life has no meaning since we are an evolved result of the existence of the universe.

Life has no meaning. Each of us has meaning and we bring it to life. It is a waste to be asking the question when you are the answer.
Joseph Campbell
(1904 – 1987)

As far as we can tell from a purely scientific viewpoint, human life has absolutely no meaning. Humans are the outcome of blind evolutionary processes that operate without goal or purpose. Our actions are not part of some divine cosmic plan, and if planet earth were to blow up tomorrow morning, the universe would probably keep going about its business as usual.
Yuval Noah Harari
(1976 –)

Another major body of thought is that life has meaning, but that its definition is up to each of us to determine for ourselves.

There is not one big cosmic meaning for all; there is only the meaning we each give to our life; and individual meaning, and individual plot, like an individual novel, a book for each person.
Anais Nin
(1903 – 1977)

Whatever we are, whatever we make of ourselves, is all we will ever have—and that, in its profound simplicity, is the meaning of life.
Philip Appleman
(1926 – 2020)

I believe that I am not responsible for the meaningfulness or meaninglessness of life, but that I am responsible for what I do with the life I've got.
Hermann Hesse
(1877 – 1962)

Yet another viewpoint holds that our ability to love and to care for others is the essence of meaning.

All you need is love.
John Lennon
(1940 – 1980)

Learn to light a candle in the darkest moments of someone's life. Be the light that helps others see; it is what gives life its deepest significance.
Roy T. Bennett
(1939 – 2014)

Love is the only way to grasp another human being in the innermost core of his personality.
Viktor E. Frankl
(1905 – 1997)

Our prime purpose in this life is to help others. And if you can't help them, at least don't hurt them.
Tenzin Gyatso, The Dalai Lama
(1935 –)

I am of the opinion that my life belongs to the community, and as long as I live, it is my privilege to do for it whatever I can.
George Bernard Shaw
(1856 – 1950)

It is not our purpose to become each other; it is to recognize each other, to learn to see the other and honor him for what he is.
Herman Hesse
(1877 – 1962)

Still another path to meaning is through what we learn and what we accomplish.

The ultimate aim of the human mind, in all is efforts, is to become acquainted with truth.
Eliza Farnham
(1815 – 1864)

The true meaning of life is to plant trees, under whose shade you do not expect to sit.
Nelson Henderson
(1865 – 1943)

The man who is born with a talent which he is meant to use, finds his greatest happiness in using it.
Johann Wolfgang von Goethe
(1749 – 1832)

The meaning of life is not simply to exist, to survive, but to move ahead, to go up, to achieve, to conquer.
Arnold Schwarzenegger
(1947 –)

Beyond work and love, I would add two other ingredients that give meaning to life. First, to fulfill whatever talents we are born with. However blessed we are by fate with different abilities and strengths, we should try to develop them to the fullest, rather than allow them to atrophy and decay... Second, we should try to leave the world a better place than when we entered it. As individuals, we can make a difference, whether it is to probe the secrets of Nature, to clean up the environment and work for peace and social justice, or to nurture the inquisitive, vibrant spirit of the young by being a mentor and a guide.
Michio Kaku
(1947 –)

Following his captivity in a German World War II concentration camp and the loss of his family, Viktor Frankl concluded that the experience of suffering can bring meaning.

If there is a meaning in life at all, then there must be a meaning in suffering. Suffering is an ineradicable part of life, even as fate and death.
Viktor E. Frankl
(1905 – 1997)

Some have even suggested that the meaning of life is not a positive one.

The human race is a monotonous affair. Most people spend the greatest part of their time working in order to live, and what little freedom remains so fills them with fear that they seek out any and every means to be rid of it.
Johann Wolfgang von Goethe
(1749 – 1832)

Human beings are so destructive. I sometimes think we're a kind of plague, that will scrub the earth clean. We destroy things so well that I sometimes think, maybe that's our function. Maybe every few eons, some animal comes along that kills off the rest of the world, clears the decks, and lets evolution proceed to its next phase.
Michael Crichton
(1942 – 2008)

Finally, one of humanity's happier traits is the ability to laugh at ourselves, and a variety of light-hearted definitions of the meaning of life are worth noting.

Everyone has a purpose in life. Perhaps yours is watching television.
David Letterman
(1947 –)

Everyone must believe in something. I believe I'll have another drink.
W. C. Fields
(1880 – 1946)

Life is what happens when you're busy making other plans.
John Lennon
(1940 – 1980)

42 is the answer to the ultimate question of life, the universe, and everything.
Douglas Adams
(1952 – 2001)

Notwithstanding all of the deep thought by prophets and philosophers, for most people the meaning of life is determined by what we do as much as by what we think. First, we have to have food and shelter. If a person is broke, hungry, and homeless, they are unlikely to be thinking about larger questions. However, many have employment in a career and derive great meaning from that. In today's world, there are tens of thousands of unique careers with clearly identifiable contribution and purpose. Although few of us will be Olympic champions or world-class concert pianists, there is still honor and meaning in being reliable and conscientious in our endeavor. Perhaps we can serve the nation or teach its children or increase the GDP.

Great meaning and satisfaction can also come from avocation, rather than vocation. There are thousands of interesting things to do that do not involve a paycheck, whether its collecting things or building things or working in a volunteer group with some common goal or cheering for your team.

My own opinion, after contemplating these definitions, is that the meaning of my life is mine to think about and to define. I favor the following dimensions:

237

1. Meaning accumulates over our entire life, and it changes with time. When I was a teenager, I found meaning in playing basketball. Then I was lucky enough to fall in love with my life partner, who like me, was raised in a home with books, classical music, and sports. That has provided continuity of meaning for us. Later, I was most concerned about providing for my family and designing electronic thingamajigs. At the moment, I am writing this book. All have separate and distinct meaning.

2. Although many things add meaning, they are not all equal.

3. Meaning is not synonymous with goodness. Being the most undetectable serial killer or the best drug lord have meaning, but not of a kind that most would favor.

4. Meaning does not require self-consciousness. A life can be meaningful whether it is self-examined or not. Plato was wrong.

5. Meaning does not require permanence or immortality. The possibility that our species may become extinct or that I may cease to exist does not erase the meaning I find in my life.

6. Meaning does not require striving for excellence. There are almost eight billion of us, and with respect to any dimension of ability or accomplishment, we are on the average, well, average, and there is no shame in that. There is also no shame in not thinking about it.

7. I want my definition to recognize that all life (including plant life) has meaning. While I was preparing to write this chapter, I happened to watch an aerial view of a herd of elephants purposefully traversing the African landscape. There were about twenty in the herd with several obviously very young members. Elephant herds are matriarchal with the senior female deciding when

and where to go. They move at the pace of the slowest member, and they are careful to protect their babies from the elements. Like human babies, elephant babies are very dependent upon their mothers, and they need to be touched every few minutes for reassurance. Elephants have exceptional memories, and they remember trails and water holes for generations. They have a range of sounds which enables them to communicate over distances of several miles. They have a lifespan of up to 70 years, and they mourn their dead. They face many of the same basic questions as did the hominids, including Homo sapiens. They too must find food and water every day, they must avoid predators, and they must reproduce in order to assure the continuity of their species. Some African herds have been smart enough to become more nocturnal in order to avoid human poachers who would murder them just for their tusks. 90 percent of an elephant's DNA is the same as mine, and I would like to acknowledge the meaning of its life.

So, here is my definition of the meaning of life:

Any of my thoughts or actions that affect my life or any other life has meaning. The meaning may be large or small, positive or negative, and may be different for me than for others. The totality of the effects of my thoughts and actions is the meaning of my life.

Applying that definition, I believe that my life had meaning before I was born because I am a tiny part of the evolution of our species and because my arrival was (hopefully) positively anticipated by my family. While I am alive, my life has meaning for many reasons, to myself and to my family and others who I have affected (such as by teaching them mathematics). Being alive is a great experience for a tiny speck of supernova residue, and I treasure it. After I am alive, my life will continue to have meaning since it is well-recorded in my family tree, and perhaps because someone may read this book and derive pleasure or meaning from it.

You might enjoy formulating your own definition of the meaning of your life. I invite you to try this simple exercise.

After careful thought, complete the following sentence: I AM _____ .

By way of illustration, my own sentence goes "I am Linda's soul mate, Greg and Shelley's dad; Neil and Grant's granddad; curious to learn about everything; kind to everyone I encounter; an engineer, a manager, and a teacher; a sports fan (Go Cardinal!); and a lover of classical music."

Notice that nationality, religion, race, wealth, or gender did not enter my particular list, even though I considered them. Perhaps some of them will make yours if they are sufficiently important to you. Give it thought and see what you learn about yourself.

Chapter 20
The Future - Where There is Life, There is Hope (And Fear)

———— ∞∞∞ ————

And the logical method and form flatter longing for certainty
which is in every human mind. But certainty generally is an
illusion, and repose is not the destiny of man."
Oliver Wendell Holmes, Jr.
(1841 – 1935)

Que sera, sera.
Whatever will be, will be.
The future's not ours to see.
Que sera, sera.
Raymond Evans
(1915 – 2007)

Given our comprehension of history, it is natural to wonder about the future, both in the near and far terms. It is possible to imagine either a utopia or dystopia, and we acknowledge the possibility of either, and anything in between, without guessing which is more likely.

The coming century is the most uncertain one in human history since the Ice Age beginning 115 thousand years ago reduced Homo sapiens population to only a few thousand. There are three major reasons. First is the risk to the biosphere caused by human pollution, as we discussed in Chapter 18. Second is the possibility of societal collapse due to nuclear or biological warfare, as we discussed in Chapter 16.

The third major factor, and the focus of this chapter, is the potential of science and technology to decisively affect our world in the coming decades. As we have noted, more than 90 percent of scientists who have ever lived are alive today. The momentum they generate will likely continue to accelerate. It is impossible to predict what technologies will have the greatest impact, although we single out a few for speculation. Be aware that there is great activity in every one of the science disciplines listed in Chapter 2. We should expect technology to continue to be a two-edged sword, with both positive and negative effects. However, the potential for technological advance becomes moot if civilization regresses due to warfare or biosphere deterioration.

Finally, we review the long-term factors affecting life on this planet. Although civilization may change dramatically over a single century, we still exist in a universe that has existed for billions of years and will continue to exist for trillions more. Earth is only temporarily habitable, and so we briefly comment on what will happen to our home planet.

Artificial Intelligence and Quantum Computing

Consider this question: For the three mentally challenging games, Jeopardy, chess, and Go, what do the current champion players in the world have in common? The answer in each case is that the best player is a machine, not a human. The reason is modern computers can store more data, analyze it more quickly, evaluate more alternatives, and choose better answers than any person. Designing computer programs to win these games is an early example of what has come to be called "Artificial Intelligence" (AI for short).

AI is defined as the theory and development of computer systems able to perform tasks that normally have required human intelligence. AI is exploding as a field of study within science. For example, the MIT "Technology Review" annually publishes a roster of 35 leading scientists under the age of 35 with summaries of their field of interest. In this year's edition, 18 out of 35 have research interests focused either on AI or in the development of advanced computing concepts to provide the greater computing capacity needed for many AI applications.

The applications of AI are pervasive including diagnosing illness, optimizing teaching techniques, autonomous weaponry, market analysis and stock trading, self-driving vehicles, manufacturing processes, facial recognition, social media monitoring, music composition and performance, disease mapping and analysis, language translation, weather forecasting, strategies in athletics, and many other fields that require the acquisition and processing of data.

The potential benefits of AI are obvious and enormous, which is why so much scientific recognition is being given to it. For example, Bill Gates has said the power of AI is "so incredible that it will change society in some very deep ways." Certainly, all of those MIT-identified scientists are sincerely determined to make the world a better place.

On the other hand, AI raises deep philosophical questions as to what it means to be human. What do our lives mean when we are no longer the highest intelligence? How do we find use for all of those whose careers (doctors, warriors, teachers, truck drivers, etc. etc.) may be diminished or eliminated? What does democracy mean in a world where public opinion is effectively manipulated and propagandized by AI applications? Is it possible that the intelligence that we create may achieve consciousness and determine that we are no longer needed? Stephen Hawking said that artificial intelligence "will either be the best thing that's ever happened to us, or it will be the worst thing. If we're not careful, it very well may be the last thing."

Accompanying the emergence of AI is a new concept of computers called "Quantum Computers." Since their emergence in the 1950's, all digital computers have been built around the concept of a "bit", short for "binary digit." A bit is simply the position of an on/off switch in which the on position can be interpreted as "one" and the off position as "zero." A collection of eight bits is called a "byte" which can be encoded in any one of 256 ways, which is more than adequate to represent all of the symbols on the keyboard. Thus, digital computers, until now, have primarily been enormous quantities of simple switches, implemented by semiconductors, operating at electronic speed.

With the development of quantum mechanics, a new type of storage element has been invented, called the Qubit. Qubits are based on detecting the states of photons in atoms at temperatures near absolute zero (-273 degrees Centigrade) which can simultaneously exist in multiple states as determined by quantum theory. Thus, a qubit encodes much more information than a bit and can be sensed at subatomic speeds. A quantum computer is a collection of qubits.

Although quantum computing is still in its early stages of development, it is believed that computers up to 100 million times faster than current ones can ultimately be possible. This will enable problems to be solved which are far beyond the capability of today's digital computers, including all of the AI applications listed above.

Life Extension

The worldwide average lifespan of Homo sapiens has increased from about 25 years to over 70. The longest life on record is a French woman, Jeanne Calment, who lived 122 years from 1875 to 1997. It is surprising, perhaps, that no one has been proven to live longer, but that is the case. However, the number of people living past the age of 100 is increasing rapidly, and is now over half a million worldwide.

Research labs all over the world are seeking to understand what maximum lifespan is possible, and what techniques might enable that maximum to be reached. At the present time, there is no consensus, but with some general view that an average of at least 115 years might be possible within a century. There are a few who speculate that a lifespan of several hundred years is possible.

A number of the very wealthy are investing large sums to investigate radical techniques for life extension. Some claim that the first person who will live to be one thousand years old, by means such as organ and DNA replacement, is already alive today. The first laboratory attempt at a cranial transplant was recently made. (Pause to consider the implications of that!)

Any significant extension to lifespan will raise serious issues for humanity. If the planet is already overpopulated, why would we want people to live forever? What would be required to enable people to stay engaged in life for 115 years or more, and what would it mean for employment and careers? What if radical extension of life is only available to the very wealthy, as would seem likely? Does civilization benefit from immortality?

Nanotechnology

Nanotechnology is the development and application of materials and products whose dimensions and tolerances are in the range from 1 to 100 nanometers. A nanometer, by definition, is one billionth of a meter, which is many times smaller than the human eye can distinguish. A single human hair is about 80 thousand nanometers wide. Thus, nanotechnology is the application of extremely small things at the level of individual atoms.

Although the potential of nanotechnology was first described in 1959 by physicist Richard Feynman, modern nanotechnology began in 1981, with the invention of the scanning tunneling microscope, which can resolve individual atoms. Nanotechnology is only now becoming a rapidly growing industry, with continuing high rates expected for many years to come.

An early success for nanotechnology was graphene, first discovered in 2004. Graphene is a sheet of carbon only one atom thick. It is one of the strongest materials known, while at the same time being an excellent conductor of both heat and electricity, optically transparent, and impermeable to gases. It is both harder than diamonds and a better conductor than copper. Nanomaterials have found application in medicine, energy, textiles, automobiles and many others. Imagine a house whose paint also serves as a solar panel to generate energy and whose bricks serve as a storage battery. Tiny nanorobots can potentially be used to deliver medications to specific parts of the body, and drones about the size of flies can provide high quality surveillance capability.

Unfortunately, nanotechnology also presents unresolved environmental risks. For example, Inhaling nanoparticles, by any living organism, may cause cell and lung damage, and there may also be other toxic effects.

Genetic Engineering

The molecule now known as DNA was first discovered in Switzerland in the late 1860's. Its role in defining the genetic characteristics of species was not fully recognized until the 1940's, and its structure as a double helix was not known until 1953. Subsequently, modern molecular biology has been concerned with understanding the role that genes play in controlling the processes of living species; plants, animals, bacteria and viruses. The Human Genome Project, completed in 2013 at a cost of three billion dollars, identified that there are about 20 thousand to 30 thousand genes in the human genome with some uncertainty remaining. All humans are more than 99 percent identical in their genetic makeup, with differences in the remaining one percent providing clues to the causes of diseases.

Scientists soon became interested in discovering whether or not the genetic makeup of a species could be modified to meet specific objectives. This was first successfully accomplished in 1973 with bacteria and mice being the first subjects. The field of genetically modified organisms (GMO) grew rapidly, and GMO for many edible crops have been developed. For example, the first gene-modified soybean was introduced in 1995 and now accounts for more than half of the crop in the United States.

Since many diseases in human beings arise from defects in their DNA, the potential benefits of genetic engineering are apparent. Research in this area led to the development of a gene-editing technique, called "CRISPR-Cas9" (Clustered Regularly

Interspaced Short Palindromic Repeats - CRISPR associated protein 9 - pronounced "crisper"), accomplishing the amazing feat of being able to identify, remove, and replace specific sections of DNA. This technique has revolutionized the field of genetic engineering. Its inventors, Jennifer Doudna and Emmanuelle Charpentier, were awarded the Nobel Prize in 2020.

CRISPR offers the potential for treating genetic diseases, including cancer, muscular dystrophy, sickle cell anemia, Huntington's disease, HIV, childhood blindness, chronic pain, joint degeneration, and many others. The technique is also broadly applicable to plants and animals. Diseases that impact food crops could be removed, or characteristics that enhance their yield could be added. Many uses in editing small animal and insect genomes are being explored with goals such as eliminating Lyme's disease.

However, CRISPR carries with it serious ethical issues. While editing the DNA of an individual after birth is not heritable (that is, it cannot pass to future generations), altering the DNA of an embryo causes those alterations to be passed on to future generations, thereby permanently altering the human genome. This raises the potential that it might be possible to design alterations to DNA to create humans with specific traits such as increased intelligence, strength, lifespan, or any other attribute of interest. It can be imagined that the capabilities of artificial intelligence and quantum computing could be united with genetic engineering tools such as CRISPR to design the genome of a new kind of human being, perhaps to improve survivability in space travel, a super warrior or any other purpose that you can imagine. For the moment, most nations have discouraged or banned research in this area, but history suggests that if the capability exists, someone will seek to weaponize it. Albert Einstein could have told you about that after nuclear weapons appeared as a result of atomic particle research.

Other Key Areas in Brief

We will touch briefly on several other topics that are important in envisioning the near future, but are more evolutionary than revolutionary. These are total population, energy, food and farming, transportation, and space exploration.

Assuming no radical impacts of warfare, global warming, or pandemic, the United Nations Population Division projects that total human population will level off at about 11 billion by 2050 compared to today's population of 7.8 billion. Two-thirds of the growth is expected to occur in Africa; India will surpass China as the most populous country; and Nigeria will replace the United States as the third largest in headcount. The populations of China, Brazil, Russia, and Japan will actually be less than today. The world will continue to become more urbanized, with more than 70 percent living in urban settings in 2100, as compared to about 55 percent at present.

The total demand for energy will increase more rapidly than the population as worldwide standard of living improves. Most forecasts indicate a 50 percent increase in energy demand by 2050. The much-maligned fossil fuels; oil, coal, and natural gas; will still provide much of the total energy. However, renewable sources; solar, wind, hydropower, and geothermal; will grow from less than ten percent to more than thirty percent. Although nuclear fusion provides the hope of inexhaustible energy with low environmental impact, it is believed unlikely to provide any significant contribution to the total before late in this century.

As population increases, the demand for food will also increase. The most important crops are wheat, rice, corn, and barley, and it is estimated that we will need to expand their total yield by 85 percent by 2050 in order to feed the world's population adequately. There is a significant link between the increase in food production and the pollution and destruction of rain forests. The present system of agriculture is wasteful in several ways. About 20 percent of crops are lost in the field due to pests, and it takes about 20 calories to produce one calorie of meat.

Aeroponics and plant-based meat are two rapidly growing industries which may address these issues. Aeroponics is the process of growing plants indoors in a mist or air environment, with nutrients supplied directly to their roots. Aeroponics offers the potential of greatly increased crop yield, while requiring 90 percent less acreage and 95 percent less water than traditional farming. Moreover, aeroponics can be accomplished in high-rise facilities in the center of urban areas. A wide variety of plant-based meats is becoming available, with some forecasting that its use will surpass consumption of animal-based meat within 20 years. Kentucky Fried Chicken, Chipotle, and McDonald's have recently introduced menu items using plant-based meats

At the dawn of civilization five thousand years ago, the development of soil-based agriculture and the domestication of animals were the enablers for humans to move from hunter-gathering cultures to civilization as we know it. We can speculate that within the next millennium, soil-based agriculture and domestic animals as a source of meat could both largely vanish.

In commercial air travel, the major impending development is hypersonic aircraft, which fly at five times the speed of sound, or faster than 3800 miles per hour. Hypersonic weapons are already nearing deployment by the major powers, and preliminary designs are in place for commercial versions. First availability may be as early as 2030. Flight time from New York to London would be 90 minutes, as compared to six hours for current jet aircraft.

For automobiles, the great trend is towards electric vehicles, with some forecasting that virtually all new autos sold by 2040 will be electric. General Motors has set the goal of going all electric by 2035. A second possibility for automobiles is the common use of self-driving cars which have already been shown to be safer overall than human drivers.

The grand vision of space travel in this century is the colonization of Mars. This goal has the romantic objective to provide the first step to survivability for our species when life on Earth has ended. The less romantic objective is to determine if Mars can be an economically viable source of mining products. One proposed timeline of development between now and the 22nd century offers a goal of a population on Mars of 30 million people in an independent nation-state. The obstacles to these goals are enormous. The cost of the project would be extraordinary. We have not proven that humans can survive, physically and emotionally, in an extraterrestrial environment. Transforming Mars into a habitat suitable for humans may prove impossible. However, the idea of extending the reach of humanity into space will retain its hold on our imagination, even if it proves to be a pipe dream.

The Future of Earth and Life Upon It

Our dear Mother Earth, like all of us, faces the certainty of her end. At best, Earth will exist for about five to seven billion more years until the Sun reaches the end of its main sequence and becomes a red giant, perhaps engulfing our planet.

Unfortunately, the ability of Earth to support human life will end much earlier. The most immediate danger is nuclear warfare, which could occur at any time. Another near-term danger is that our species could be one of the victims of the on-going Anthropocene extinction event which we have perpetrated. This could occur within a few centuries. Destruction of the biosphere via pollution or climate change is a subset of this possibility.

Major volcanic eruptions occur about once in every forty years, and they are occasionally enough to impact weather globally. If humans survive long enough, we will have to cope with ice ages which occur every 100 thousand to 400 thousand years. Brief periods of global warming will occur at similar intervals with major effect on sea level. The Earth's magnetic

Artificial intelligence, problem solving capability by machines, is advancing rapidly and being applied to many areas such as medical diagnoses. A concern is what will it mean to civilization if humans are no longer the highest form of intelligence.

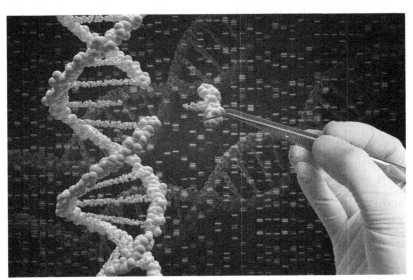

Techniques such as CRISPR-Cas9 are enabling the editing of the human genome. This offers the potential to treat genetic diseases such as muscular dystrophy, but raises ethical questions concerning modifying the human genome to create specialized humans.

Aeroponics is the process of growing plants indoors without the use of soil. Crops can be grown year-round in a high-rise urban environment with greater yields from less water and space than traditional farming.

poles reverse at random intervals ranging from 10 thousand years to 50 million years. This can potentially make us more vulnerable to solar radiation.

Human survival on Earth, at most, will be from half a billion to a billion years, limited by the increasing brightness of the Sun which will ultimately destroy the atmosphere. Many other cosmological disasters are possible including collision with a large asteroid (such as the one that wiped out the dinosaurs a mere 70 million years ago), a supernova within 100 light years of Earth, solar flares, and gamma ray bursts.

Nevertheless, it is imaginable that humanity will have time and ability to populate the solar system, then extend into the Milky Way, then to galaxies beyond, and then to other universes trillions of years into the future. As the Roman orator, Cicero (106 BCE – 43 BCE) first said so long ago, and Stephen Hawking repeated in 2018, "While there's life, there's hope."

Chapter 21
Ideas - From the Trivial to the Colossal

Everything begins with an Idea.
Earl Nightingale
(1921 – 1989)

One of the saddest lessons of history is this: If we've been bamboozled long enough, we tend to reject any evidence of the bamboozle. We're no longer interested in finding out the truth.
Carl Sagan
(1934 – 1996)

This chapter has one fundamental thesis: Most of the defining characteristics of our civilizations have their origin as an idea that began in the mind of a single individual and grew from that mind to affect billions of people. There are relatively few ideas with such impact. Perhaps the ones with the most impact on humanity are these seven (ordered by when they occurred, with the originator and date of origin):

Monotheism – Abraham (?), Sumeria, ca. 1900 BCE

Reincarnation – Unknown originator, Indus River Valley, ca. 800 BCE

Natural Rights – John Locke, England, 1689

Capitalism – Adam Smith, Scotland, 1776

Evolution of Species – Charles Darwin, England, 1831

Marxism – Karl Marx, Germany, 1847

The Big Bang and Expanding Universe – Edwin Hubble, United States, 1929

An idea is defined as "a thought as to a possible course of action." We have ideas all the time, including while we are asleep. One recent study estimates that we each have over six thousand thoughts per day. Thus, in an 80-year lifetime, each of us has over 150 million thoughts that cross our minds. Most are not very important, for example "I think I will tie my shoes." Others might carry more import, like "I think I will run for President."

One of humanity's most remarkable capabilities is that there is no limit to our imagination or the ideas we can conceive. All fiction written over the centuries is testimony to this. We can imagine Big Bangs and expanding universes. We can imagine an omnipotent God who offers us immortality. Recently, some have imagined that there are infinitely many universes where all alternative laws of physics exist and every imaginable circumstance occurs. Another recent idea is that we are merely fictions in some higher power's giant software program and that we do not really exist at all.

One of our skills is that, if we have an idea we believe is good, we may set about to convince others of its merit. We might have many motivations to do this: the belief that we have received divine revelation; to make a contribution to the world; to project a political viewpoint; to profit from our idea; or in the worst case, the conscious motive to cheat or deceive. We might not even fully understand our own motivation.

History repeats many examples of one charismatic mind gaining a following and leading others to disaster and death. Just as we have the wonderful ability to imagine anything, humans have the less wonderful attribute of being able to be induced to believe the absurd, given effective persuasion. We can believe that we should burn witches at the stake, or rid the world of ethnic or racial groups or whomever we dislike. The Flat Earth Society has many committed members. The examples are endless.

Unfortunately, sometimes believing the absurd can be fatal. In 1974, Marshall Applewhite and Bonnie Nettles convinced themselves that they had been chosen to fulfill a biblical prophecy of dying, being restored to life, and transported to a spaceship. They recruited 38 followers over 23 years who agreed to commit ritual suicide in the belief that their souls would be transported to a spaceship reputed to be following the Hale-Bopp Comet, and then a UFO would take them to another level of existence. They all faithfully committed suicide in March, 1997.

Although this may seem like an isolated and bizarre case of gullibility, contemporary communications and social media provide an effective and nearly limitless capacity to promote disinformation, lies, and half-truth in a desire to manipulate public opinion towards the political, religious, or financial goals of the purveyor.

The beliefs promoted by today's Q-Anon conspiracy provide a current example. Q-Anon is promoting, with support from elected politicians and broadcast media, that a cabal of thousands of high-ranking members of the government and other public figures are Satanic, cannibalistic pedophiles operating a global sex-trafficking ring. According to the most recent polls, about 20 percent of American voters believe some or all of the Q-Anon story.

These examples prove that, no matter how implausible an idea, it can find followers who will support the idea with votes, money, and even their lives.

Moving away from the absurd and perverse, let us return to the most influential ideas that have survived and molded our civilization. There is a process that all have in common, beginning with origin in one mind, ultimately to be adopted in the minds of the billions who follow hundreds or thousands of years later.

1. The originator of the idea must present his case convincingly, in the form of spoken or written words.

2. Initial advocates must appear who firmly believe the idea and further promote it to others. Supporting books and testimonials are written. The founding idea may be modified and extended to support changing circumstances.

3. As the number of advocates increases, an infrastructure develops that supports the idea and is committed to its merit. In the case of the religions, dedicated structures (e.g., cathedrals and mosques) are built and

a profession of clerics develops. Funding mechanisms are established. For the science and economics oriented ideas, books are written, experiments done, university courses established, and related professions (such as Professor of Biology or Economics) appear. Adherents become committed to the idea and resistant to alternatives, since they have dedicated their lives to the idea. The idea may continue to evolve. Perhaps different branches may develop. For example, Judaism led to Christianity which led to Islam, and in all three there are many distinct sects. All, widely divergent, are within the framework of monotheism and dedication to the worship of one Deity. Similarly, Marxism as developed in Russia by Lenin and Stalin and then in China by Mao Tse Tung were widely divergent from each other and from Marx's original concept.

4. As the power of the advocates increases, it becomes a requirement that the idea be accepted as a matter of law. Those who do not accept the idea may be ridiculed, persecuted, exiled, imprisoned, tortured, or murdered. This has been especially true for the Christian and Muslim faiths as well as for Capitalism and Communism. To date, the Big Bang and evolution ideas have not been adopted into law by any government.

5. After nations and their leaders have embraced the idea, they become willing to go to war to defend and expand it. In the period after the fall of Rome through the Renaissance, Europe and the Middle East were consumed with nearly continuous religious warfare between Christianity and Islam, between Catholics and Protestants, and between Sunni and Shia Muslims. Then, in the eighteenth and nineteenth centuries, numerous revolutions were fought to obtain the rights of man as first proposed by John Locke. Finally, the

period after World War II has been largely defined by the struggle between capitalism and communism.

The following seven ideas, each emerging from the mind of a single individual, have had enormous impact on the lives of the billions of people who have lived since the first civilizations only six thousand years ago. (It is likely that each of these individuals was influenced by earlier ideas, but nevertheless history has identified them as "the father of" the idea in question.)

Monotheism – Although the prophet Abraham may be mythical, he is a key figure in Judaism, Christianity, and Islam. About 1900 BCE, someone, believed to be Abraham, first articulated that there is a single all-powerful and all-knowing God who demands faith as a condition for salvation. The idea of monotheism can be stated as "the belief in one God who created the world and is omnipotent, omnipresent, omnibenevolent, and omniscient." Today, more than 55 percent of the world's population, or more than 4 billion people, are adherents of one of the Abrahamic religions, and this number is expected to continue to grow in coming decades.

Reincarnation – The idea of reincarnation was first expressed about 800 BCE. Although no single person is named to this idea, there had to be one. This idea, and the accompanying near-endless cycle of life, unites the Hindu and Buddhist faiths. It can be stated as "once a living being dies, its atman (soul) will be reborn or reincarnated into a different body depending on its karma from its previous life." Today, there are over 1.5 billion people in the world who are Hindus or Buddhists, or about 20 percent of the world's people.

Natural Rights – John Locke was an English philosopher and political theorist who is often called "the father of liberalism." In 1688, he was the first to articulate that there are three fundamental rights of mankind; life, liberty, and property, and that the purpose of government is to protect those rights. Thus, he anticipated and helped to inspire the English Bill of Rights, the American Declaration of Independence, the French Revolution, and others. The Universal Declaration of Human Rights, signed in 1948 by all 193 members of the United Nations, incorporates Locke's ideas, although not nearly all of those nations are in full compliance or agreement. Today, the British magazine, The Economist, maintains a Democracy Index which measures the state of democracy in the world. By their most recent analysis, about 50 percent of the world's population (i.e., about 4.4 billion people) lives in either full democracies or flawed democracies, and 50 percent live either under fully authoritarian regimes or regimes in which elections are held, but are fraudulent and predetermined. They rank Australia, New Zealand, and the Netherlands as the most democratic countries.

Capitalism – Adam Smith was a Scottish economist who is considered the father of modern economics and capitalism. In his 1776 book, "An Inquiry into the Nature and Causes of the Wealth of Nations," he proposed that people work for their own good, that competition forces people to make better products at lower cost, and that the law of supply and demand results in the most product at the lowest cost. He proposed that a laissez-faire ("hands off") policy of minimum intervention of government in the regulation of economy and competition is best. The debates of capitalism versus communism and capitalism versus socialism are the most aggressive and emotional of our time, often with much heat and little light. While it seems clear that communism has failed, it is less clear that capitalism has succeeded, at least in terms of protecting the environment and enabling a reasonable distribution of wealth. Nevertheless, capitalism has affected every person on the planet, and in that sense, exceeds all of the other six key ideas presented here.

Evolution – Charles Darwin was an English naturalist with the idea that all species of life evolved by the process called natural selection. He published his seminal work, "On the Origin of Species" in 1859, although he had first conceived the idea in 1838. Natural selection occurs due to changes in the environment and in an individual organism due to random variation. Fellow English naturalist, Alfred Russel Wallace, also conceived the idea of natural selection shortly after Darwin in approximately 1856, although both shared credit for the discovery via joint publication. The principles of evolution are now broadly accepted throughout the world by about 65 percent of all people and about 97 percent of scientists, according to a 2021 Pew Research study. The rate of acceptance in Europe and Asia is about 80 percent, but that falls to 50 percent or less in countries, including the United States and South America, where religious fundamentalism remains strong.

Marxism – Karl Marx was a German philosopher and economist who, together with Friedrich Engels published the Communist Manifesto in 1847. By then, the Industrial Revolution had yielded major industries driven by capitalist principles, including acquiring labor at the lowest cost and maximizing the profit of the capitalist. This created draconian practices such as the employment of children as laborers in mines and textile mills. Marx predicted that this would inevitably lead to permanent class conflict, and his proposed response was the abolition of private property and public ownership of the means of production. His ideas were seized by Vladimir Lenin, a Russian, who favored a dictatorship of the working class, to be accomplished via revolutionary action. This led to the overthrow of the Russian Tsar Nicholas in 1917 and the establishment of the dictatorships of Lenin and Josef Stalin. Following World War II, Russia extended its empire into the Union of Soviet Socialist Republics (USSR), and China followed into the communist orbit in 1949 under Mao Tse Tung. The communist countries were aggressively expansionist, forcing strong resistance from other countries, led by the United States. Ultimately, the USSR collapsed in 1991 and modern China remains governed by its Communist party, although its economic behavior is more consistent with capitalism. Although Marxism is a failed ideology, it has had strong effect on the lives of billions of people, albeit negative.

The Big Bang – The American astronomer, Edwin Hubble, observed in 1926 that every galaxy he observed is receding from the Earth. This led to the proposal by Georges Lemaitre, a Belgian cosmologist, in 1927 that the universe began with a single gigantic explosion, later named (in 1949) the "Big Bang" by English astronomer, Fred Hoyle. Although there is still debate about the earliest phases of the Big Bang, the idea of a giant initial explosion is accepted by virtually all astronomers today. A major consequence of the Big Bang and Darwin's principles of evolution is to remove Homo sapiens from the center of creation to the periphery.

The seven key ideas have not only strongly influenced their adherents, but also their opponents, as a result of the energy and resources consumed in resisting. For example, the negative impact of Karl Marx's idea on the United States following World War II is difficult to overstate. He is reviled. Without him, the Cold War with the Soviet Union would not have occurred, or the "loss" of China, the Korean War, the Vietnamese War, or the Cuban missile crisis. Instead of spending more on weapons than the rest of the world combined and maintaining a world-wide network of military bases, the United States might have invested more in promoting the ideas of Thomas Jefferson, at home and abroad, than in resisting those of Karl Marx. The current divided state of the nation might have been averted. To this day, the worst pejorative one American can call another is to label him a communist or Marxist, even worse than a Fascist or Nazi. Should the American republic fall to authoritarianism, it could be argued that the proximate cause was the idea of Karl Marx.

Of our seven decisive ideas, two originated thousands of years ago and five originated within the last several hundred. Others may appear as humanity moves forward. Perhaps there is a way for the world to govern itself with less conflict among nations. Perhaps it is possible to more equitably share the wealth of the planet while still enabling creative initiative and progress. Perhaps there is a way to assure the continuing habitability of the planet. Perhaps the next great idea is already in some single mind.

Chapter 22
A Summary in Quotations –
Your Conversation with the Ages

—————∞∞∞—————

Throughout my life, now spanning eight decades, I have enjoyed the pungent one-line quote as a way of communicating more than the sum of the words. My father was fond of saying "a prophet is without honor in his own bailiwick." My two favorites are "the price of achievement is toil, and the gods have ordained that you pay in advance" (first thought to be uttered by the Greek poet, Hesiod, in the eighth century BCE) and "certainty is an illusion and repose is not the destiny of man", a slight variation of words said by Oliver Wendell Holmes early in the twentieth century. I have repeated these ad nauseum to co-workers, students, friends, spouse, children, and grandchildren for my amusement and their edification and annoyance.

In planning the organization of this book, I set the parameter of beginning each chapter with two quotes selected to illuminate its principal ideas. In choosing them, I reviewed thousands of candidates before narrowing the choice to two per chapter. In most cases, I found far more than I could use, and to my regret, left out many great thoughts. It occurred to me that these semi-finalists are worth sharing, so I present them here, without further comment, as a form of summary.

I have organized them by relevance to a particular chapter, providing linkage to the rest of the book. You may find a few that inspire further thought or even incorporation into your own lexicon. I hope that you find some to enlighten, challenge, or enjoy. I particularly recommend the lengthy quote by English art historian, Kenneth Clark, listed under Chapter 11.

These quotes are not here necessarily because I agree with them, but because they are challenging, provocative, interesting, or in some cases, funny. With these thoughts, you are in conversation with past civilizations and some of the greatest minds from the time of the first written word to this very moment.

PART I – COMPREHENDING YOURSELF

Chapter 1 – Allow Me to Introduce Yourself: Comprehension Begins at Home

We are just an advanced breed of monkeys on a minor planet of a very average star. But we can understand the universe. That makes us something very special.
Stephen Hawking
(1942 – 2018)

Look at everything as though you were seeing it for the first time or the last time. Then your time on Earth will be filled with glory.
Betty Smith
(1896 – 1972)

You only live once, but if you do it right, once is enough.
Mae West
(1893 – 1980)

Chapter 2 – How We Know Things: Science and the Scientific Method

Wisdom is knowing that you don't know.
Socrates
(469 BCE – 399 BCE)

When the scientific method came into being, it gave us a new window on the truth; namely a method by laboratory-controlled experiments to winnow true hypotheses from false ones.
Huston Smith
(1919 – 2016)

This is one of the most important lessons of the scientific method: if you cannot fail, you cannot learn.
Eric Ries
(1978 –)

Science doesn't make it impossible to believe in God. It just makes it possible not to believe in God.
Steven Weinberg
(1933 – 2021)

The best thing about science is that it's true whether you believe it or not.
Neil deGrasse Tyson
(1958 –)

Truth is incontrovertible, ignorance can deride it, malice may destroy it, but there it is.
Winston Churchill
(1874 – 1965)

Chapter 3 – Numbers Large and Small: Learning How to Count

Man is the measure of all things.
Protagoras
(480 BCE – 411 BCE)

When you can measure what you are speaking about, and express it in numbers, you know something about it, but when you cannot measure it, when you cannot express it in numbers, your knowledge is of a meagre and unsatisfactory kind.
Lord Kelvin
(1824 – 1907)

Chapter 4 – Roadmap: Milestones of Your Universe and Your World

Why is there something rather than nothing?
Gottfried Wilhelm Leibniz
(1646 – 1716)

I can calculate the motion of heavenly bodies, but not the madness of people.
Isaac Newton
(1643 – 1727)

*Two things are infinite – the universe and human stupidity, and
I'm not sure about the universe.*
Albert Einstein
(1879 – 1955)

PART 2 - COMPREHENDING YOUR UNIVERSE

Chapter 5 – The Universe: BANG! You Are Alive

*Had I been present at the Creation, I would have given some
useful hints for a better ordering of the universe.*
Alfonso the Wise
(1221 – 1284)

*String theory has the potential to show that all of the wondrous
happenings in the universe –- from the frantic dance of
subatomic quarks to the stately waltz of orbiting binary stars;
from the primordial fireball of the big bang to the majestic swirl
of heavenly galaxies – are reflections of one, grand physical
principle, one master equation.*
Brian Greene
(1963 –)

*If the rate of expansion one second after the Big Bang had been
smaller by even one part in a hundred thousand million million,
it would have recollapsed before it reached its present size.
On the other hand, if it had been greater by a part in a million,
the universe would have expanded too rapidly for stars
and planets to form.*
Stephen Hawking
(1942 – 2018)

Chapter 6 – The Milky Way: A Galaxy of Stars

*The Milky Way is nothing else but a mass of innumerable
stars planted together in clusters.*
Galileo Galilei
(1564 – 1642)

*We live in a modest system, a galaxy called the Milky Way. If
we named every star in the Milky Way and put them in the
Hollywood telephone directory and stacked those telephone
directories up, we'd have a pile of telephone directories
70 miles high.*
John Rhys-Davies
(1944 –)

Chapter 7 – The Sun: The Light of Your Life

The Sun is the giver of life.
Ramses II
(1279 BCE -1213 BCE)

If I had to have a religion, I should adore the Sun, for it is the Sun that fertilizes everything.
Napoleon Bonaparte
(1769 – 1821)

Chapter 8 – The Earth: Your Home

Earth provides enough to satisfy every man's needs, but not every man's greed.
Mahatma Gandhi
(1869 – 1948)

It suddenly struck me that that tiny pea, pretty and blue, was the Earth... . I felt very, very small.
Neil Armstrong
(1930 – 2012)

Far out in the uncharted backwaters of the unfashionable end of the western spiral arm of the Galaxy lies a small unregarded yellow sun. Orbiting this at a distance of roughly ninety-two million miles is an utterly insignificant little blue green planet whose ape-descended life forms are so amazingly primitive that they still think digital watches are a pretty neat idea.
Douglas Adams
(1952 – 2001)

What is the use of a house if you haven't got a tolerable planet to put it on?
Henry David Thoreau
(1817 – 1862)

The earth will not continue to offer its harvest, except with faithful stewardship. We cannot say we love the land and then take steps to destroy it.
Pope John Paul II
(1920 – 2005)

Chapter 9 – Life on Earth: From One Cell to the Next

*We are the representatives of the cosmos; we are an
example of what hydrogen atoms can do,
given 15 billion years of cosmic evolution.*
Carl Sagan
(1934 – 1996)

The love for all creatures is the most noble attribute of man.
Charles Darwin
(1809 – 1882)

*Evolution is the fundamental idea in all of life science –
in all of biology.*
Bill Nye
(1955 –)

*The human mind evolved to believe in the gods. It did not evolve
to believe in biology.*
E. O. Wilson
(1929 – 2021)

*Theology made no provision for evolution. The biblical authors
had missed the most important revelation of all! Could it be that
they were not really privy to the thoughts of God?*
E. O. Wilson
(1929 – 2021)

Chapter 10 – Homo Sapiens: The Organism

*The human brain has 100 billion neurons, each neuron
connected to 10 thousand other neurons. Sitting on your
shoulders is the most complicated object in the known
universe.*
Michio Kaku
(1947 –)

*From the dust of the earth, from the common elementary fund,
the Creator has made Homo sapiens. From the same material
he has made every other creature, however noxious and
insignificant to us. They are earth-born companions and our
fellow mortals.*
John Muir
(1838 – 1914)

267

PART 3 - COMPREHENDING YOUR WORLD

Chapter 11 – Preview of Part 3: What Will Be, Will Be

I believe order is better than chaos, creation better than destruction. I prefer gentleness to violence, forgiveness to vendetta. On the whole I think that knowledge is preferable to ignorance, and I am sure that human sympathy is more valuable than ideology. I believe that in spite of the recent triumphs of science, men haven't changed much in the last two thousand years; and in consequence we must try to learn from history.
Kenneth Clark
(1914 – 2005)

If I had to choose between an erudite Aristotle and an unknown 'soulless' black slave I would choose the latter. The ascendancy of the West was on a heap of bodies of slaves and trampled humanity through colonization.
Viktor Vijay Kumar
(unknown)

Chapter 12 – Hunter/Gatherers: Down from the Trees

No arts; no letters; no society; and which is worst of all, continual fear and danger of violent death; and the life of man solitary, poor, nasty, brutish, and short.
Thomas Hobbes
(1588 – 1679)

Early humans, bursting with questions about Nature but with limited understanding of its dynamics, explained things in terms of supernatural persons and person-animals who delivered the droughts and floods and plagues.
Ursula Goodenough
(1943 –)

Chapter 13 – Civilization: From the Farm to the Empire

A great civilization is not conquered from without until it has destroyed itself from within.
Ariel Durant
(1898 – 1981)

If a nation expects to be ignorant and free, in a state of civilization, it expects what never was and never will be.
Thomas Jefferson
(1743 – 1826)

Civilization exists by geological consent,
subject to change without notice.
Will Durant
(1885 – 1981)

Chapter 14 – Feudalism to Domination: Europe Conquers the World

Feudalism made land the measure and the master of all things.
Lord Acton
(1834 CE – 1902 CE)

Not much was really invented during the Renaissance, if you
don't count modern civilization.
P. J. O'Rourke
(1947 – 2022)

Learning never exhausts the mind.
Leonardo da Vinci
(1452 – 1519)

I may disagree with what you have to say, but I shall defend, to
the death, your right to say it.
Francois-Marie Arouet (Voltaire)
(1694 – 1778)

Chapter 15 – Modern Times: Revolutions and Ideologies

In the Industrial Revolution, Britain led the world in advances
that enabled mass production: trade exchanges, transportation,
factory technology, and new skills needed for the new
industrialised world.
Lucy Powell
(1974 –)

Our great democracies still tend to think that a stupid man is
more likely to be honest than a clever man, and our politicians
take advantage of this prejudice by pretending to be even more
stupid than nature made them.
Bertrand Russell
(1872 – 1970)

How do you tell a Communist? Well, it's someone who reads
Marx and Lenin. And how do you tell an anti-Communist? It's
someone who understands Marx and Lenin.
Ronald Reagan
(1911 – 2004)

*Globalization is a fact, because of technology, because of
an integrated global supply chain, because of changes in
transportation. And we're not going to be able to build
a wall around that.*
Barack Obama
(1961 –)

Chapter 16 – Warfare – BANG! You Are Dead

It is well that war is so terrible, or we should grow too fond of it.
Robert E. Lee
(1807 – 1870)

*Almost all wars, perhaps all, are trade wars connected with
some material interest. They are always disguised as sacred
wars, made in the name of God, or civilization or progress. But
all of them, or almost all of the wars, have been trade wars.*
Eduardo Galeano
(1940 – 2015)

*We live in the dark ages. If an intelligent society can destroy
itself in large numbers and places the largest amount of
revenues in instruments of destruction, it is certainly not an
evolved society or an intelligent society.*
Frederick Lindemann
(1846 – 1931)

*I know not with what weapons World War III will be fought, but
World War IV will be fought with sticks and stones.*
Albert Einstein
(1879 – 1955)

*And we know for certain that some lovely day someone will set
the spark off and we will all be blown away*
Tom Lehrer
(1928 –)

In war, the first casualty is truth.
Terry Hayes
(1951 –)

*Religion makes people kill each other. Science supplies them
with weapons.*
Mokokoma Mokhonoana
(unknown)

All Welsh knew was that he was scared shitless, and at the same time was afflicted with a choking gorge of anger that any social coercion existed in the world which could force him to be here.
James Jones
(The Thin Red Line)
(1921 – 1977)

Chapter 17 – Systems of Belief: Everyone is in the Minority

My religion is very simple. My religion is kindness.
Dalai Lama
(1935 –)

The World is my country, all mankind are my brethren, and to do good is my religion.
Thomas Paine
(1737 – 1809)

Religion is regarded by the common people as true, by the wise as false, and by the rulers as useful.
Lucius Annaeus Seneca
(4 BCE? – 65 CE)

Chapter 18 – The Biosphere: To Be or Not to Be

We have found the enemy, and it is us.
Walt Kelly
(1913 – 1973)

A nation that destroys its soils destroys itself. Forests are the lungs of our land, purifying the air and giving fresh strength to our people.
Franklin D. Roosevelt
(1882 – 1945)

For the first time in the history of the world, every human being is now subjected to contact with dangerous chemicals, from the moment of conception until death.
Rachel Carson
(1907 – 1964)

PART 4 - FROM COMPREHENSION TO CONJECTURE AND CONCLUSION

Chapter 19 – The Meaning of Life: It is Your Choice

Look at everything always as though you were seeing it either for the first or last time: Thus is your time on earth filled with glory.
Betty Smith
(1896 – 1972)

I alone cannot change the world, but I can cast a stone across the water to create many ripples.
Mother Teresa
(1910 – 1997)

Without music life would be impossible.
Friedrich Nietzsche
(1844 – 1900)

I am only one, but I am one. I cannot do everything, but I can do something. And I will not let what I cannot do interfere with what I can do.
Edward Everett Hale
(1822 – 1909)

Never doubt that a small group of thoughtful, committed citizens can change the world. Indeed, it is the only thing that ever has.
Margaret Mead
(1901 – 1978)

How can the dead be truly dead when they still live in the souls of those who are left behind?
Carson McCullers
(1917 – 1967)

Chapter 20 – The Future: Where There is Life, There is Hope (and Fear)

Probably every generation sees itself as charged with remaking the world. Mine, however, knows that it will not remake the world. But its task is perhaps even greater, for it consists in keeping the world from destroying itself.
Albert Camus
(1913 – 1960)

Forests and meat animals compete for the same land. The prodigious appetite of the affluent nations for meat means that agribusiness can pay more than those who want to preserve or restore the forest. We are, quite literally, gambling with the future of our planet – for the sake of hamburgers.
Peter Singer
(1946 –)

The greatest danger to our future is apathy.
Jane Goodall
(1934 –)

Technology – with all its promise and potential – has gotten so far beyond human control that its threatening the future of humankind.
Kim Vicente
(1963 –)

Chapter 21 – Ideas: From the Trivial to the Colossal

Forgive them, Father, for they know not what they do.
The Bible
Luke 23:34

Ideas shape the course of history.
John Maynard Keynes
(1883 – 1946)

Changes in the structure of society are not brought about solely by massive engines of doctrine. The first flash of insight which persuades human beings to change their basic assumptions is usually contained in a few phrases.
Kenneth Clark
(1914 – 2005)

You can fool all of the people some of the time and some of the people all the time, but you cannot fool all the people all the time.
Abraham Lincoln
(1809 – 1865)

One of the handicaps of stupidity is its inability to imagine intelligence.
Donna Leon
(1942 –)

There is no cure for stupid.
Pamela Clare
(unknown)

He who seeks to deceive will always find someone who will
allow himself to be deceived.
Niccolo Machiavelli
(1469 – 1527)

If you tell a lie long enough, it becomes the truth.
Joseph Goebbels
(1897 – 1945)

The problem is that the people with the most ridiculous ideas
are always the people who are most certain of them.
Bill Maher
(1956 –)

A man may die, nations may rise and fall, but an idea lives on.
Ideas have endurance without death.
John F. Kennedy
(1917 – 1963)

Chapter 23 – Conclusion: Summary of the Sum and Substance

Don't worry. Be happy.
Bobby McFerrin
(1950 –)

If you want to be happy, do not dwell in the past, do not worry
about the future, focus on living fully in the present.
Roy T. Bennett
(1939 – 2014)

Chapter 23
Conclusion - Summary of the Sum and Substance

———— ∞∞∞ ————

*Today I consider myself the luckiest man on the
face of the earth.*
Lou Gehrig
(1903 – 1941)

Always look on the bright side of life.
Monty Python
("Life of Brian," 1979)

Congratulations again, dear reader! You made it to the end of this adventure in comprehension. Perhaps you learned something new, encountered a thought that was worth exploring, or found a topic to investigate further.

While you spent eight hours reading this book, a lot has happened to your body, your universe, and your world. Your body was busy producing over 25 billion new cells in those eight hours at the rate of 60 million per minute to replace the ones that died. Earth rotated on its axis by eight thousand miles (at the equator), and in its orbit of the Sun by over 500 thousand miles. The Sun rotated on its axis about the Milky Way galaxy by 3.3 million miles. Finally, the galaxy is part of a universe which is expanding at nearly the speed of light, so in eight hours you are a little more than five billion miles in space from where you started.

A lot happened here on Earth during those eight hours as well. Mother Earth welcomed about 130 thousand brand new babies, and said farewell to about 52 thousand souls, for a net growth of almost 80 thousand. All 7.8 billion of us have (hopefully) had a meal or a good night's sleep and lived a total of 62 billion hours. To provide those meals, we have slaughtered 226 thousand cattle and 45 million chickens, not to mention lots of goats and sheep. Since average global life expectancy is 72.7 years, the equivalent of 92 thousand total lifetimes were lived while you were reading. Everyone has produced more than half a pound of trash for a total of 5.2 billion pounds in eight hours, amounting to over 4 million cubic yards of landfill.

Following is a chapter-by-chapter summary which captures the sum and substance of the sum and substance.

Allow Me to Introduce Yourself – At the instant of your conception, you were only one of forty quadrillion possible pairs of DNA from your parents, so your life began with an improbable stroke of luck. Your parents determined your race, sex, language, place of residence, social standing, and economic status. The language you speak and your religion were most probably determined by whoever last conquered the place where you live. The land you temporarily own once was part of an earlier civilization and, before that, was buried in ice or covered with water. You are lucky to be alive today in this era of jet travel, large screen TV, and the smartphone.

How We Know – The scientific method emerged in the 16th century as an effective means for humanity to extend its knowledge of reality, and, through the resulting technological advances, to improve man's control of his environment. The scientific method requires a clear statement of hypothesis, the rigorous gathering and documentation of experimental results, and independent verification. The development of modern tools, such as telescopes, microscopes, clocks, scales, and many others have extended natural human senses by orders of magnitude. The continuing extension of knowledge today largely arises from the more than 25 thousand universities in the world. More than 90 percent of all scientists who have ever lived are still alive today.

Numbers Large and Small – The universe involves many numbers which are incomprehensibly large and others which are tinier than we can conveniently imagine. There are trillions and trillions of stars, galaxies, planets, molecules, cells, atoms, etc. etc. The intervals of time and distances involved at the subatomic particle level are tiny, far beneath our ability to sense them directly. Exponential notation allows us to measure and analyze these quantities without requiring large numbers of zeroes, either before or after the decimal point. The advice is offered that when you encounter numbers like 7.0×10^{27} (the number of atoms in a human being) or 6.626×10^{-34} (Planck's constant having to do with photon energy), you just observe to yourself that those are really, really big or really, really little, and move on.

Roadmap – Although there is much that is still not well understood about the basic science of the universe and its composition, current belief is that the universe as we know it began 13.7 billion years ago in a monstrous explosion called the Big Bang. Only 200 million years later, our Milky Way galaxy formed. About nine billion years later, the Sun and Earth came to exist, but Earth was initially inhospitable to life. A billion years or so later, oceans and continents appeared, and the first single-cell life form emerged (nobody knows how for sure). It took another three billion years for multiple cell life based on DNA to evolve. In another half a billion years, our first hominid ancestor appeared. Then, in another five million years, the very first modern human (Homo sapiens) appeared. This was only 250,000 years ago. The pace of things then accelerated, at least with regards to Homo sapiens. The first civilization (including written language and agriculture) emerged only six thousand years ago, followed quickly by the Scientific Revolution only six or seven hundred years ago. Then, at long last, here we are today.

The Universe – There are three independent verifications that the universe started with the Big Bang: 1) the universe was shown by Hubble and Lemaitre to be expanding from a single point of origin; 2) background radiation has been found, as predicted by the initial conditions of heat and energy of the Big Bang; and 3) the chemical composition of stars, being primarily hydrogen and helium, is as predicted by modern quantum physics. After a few hundred million years, the hydrogen and helium began to aggregate into masses large enough to form the first stars. The first groups of stars, called galaxies, formed between 200 million to 700 million years after the Big Bang. Stars generate light and energy by fusion of hydrogen into helium, and they ultimately burn out when the supply of hydrogen for fusion is exhausted. Stars come in a large variety of sizes, temperatures, and colors. The total life of a star can range from a few million years to many billions of years. If a star is massive enough, at the end of its life, it will explode into a supernova, which results in forming and disbursing elements heavier than hydrogen and helium throughout a huge region of its galaxy.

The Milky Way – Currently, it is estimated that there are about two trillion galaxies in the universe. Our home galaxy, the Milky Way, was formed only 200 million years after the Big Bang. All of the stars which can be seen with the naked eye from Earth are in the Milky Way. It is in the form of a spiral with four main arms. At the center of the Milky Way is a supermassive black hole, called so because its mass density is so great not even light can escape it. Our galaxy is rotating at the rate of once every 240 million years, so that it has rotated less than one degree since humans first appeared on Earth. The Milky Way includes over 100 billion stars, of which our Sun is one. No extraterrestrial life has yet been discovered, but there are as many as 300 million planets in the Milky Way alone and trillions in the entire universe.

The Sun – Our Sun is located about 26,000 light years from the center of the Milky Way and 93 million miles from Earth. All life on Earth depends on the energy provided by the Sun, beginning with photosynthesis of plant life, which ultimately provides food for all other life forms, including Homo sapiens. The Sun is not a solid object, and it consists primarily of hydrogen and helium. It generates energy through fusion of hydrogen at the rate of 600 million tons per second. Energy emitted from the Sun takes only eight minutes or so to reach the Earth. The Sun has a strong magnetic field, and occasional variations produce solar flares, and coronal mass ejections, which can severely damage modern satellites and other electronics. Luckily, our Sun is still relatively early in its life cycle, but it will become a red giant and will engulf the Earth in about 5 billion years. However, the Sun will only sustain an oxygen-based atmosphere for about another billion years.

Earth – Our home planet has the good fortune to exist in the so-called "Goldilocks Zone" in which liquid water can exist on a year-round basis, and thus so can life. Earth was born about 4.5 billion years ago, about 100 million years after the Sun. Earth has a radius of about 4000 miles and consists of four layers: a solid inner-core of iron and nickel; a liquid outer core of the same composition as the inner core; the mantle which consists mainly of lighter elements; and a rocky outer crust whose surface is our home. Initially, the surface was molten, and there was no oxygen-based atmosphere. Earth was inhospitable to life until the first single-cell life form appeared about 4 billion years ago. Although the surface of Earth seems stable over the course of a single human lifespan, in the long run, that is not the case. The magnetic poles reverse from time to time; the rotational axis of the planet wobbles; the tectonic plates on which the continents sit are in continual slow motion; there are occasional massive volcanic eruptions and collisions with other objects in space;

and climate varies from ice ages in which most of the surface is covered with ice, to other eras in which forests grow at the poles. Nevertheless, we are fortunate to live in an era of moderate climate, and at a time when there is possibly no more beautiful place in the universe than the surface of Earth.

Life on Earth – Nobody knows for sure how or why single-cell life appeared on Earth four billion years ago, but it did. It took more than another billion years for multiple-cell organisms to evolve, and from there, the process of natural selection took off, ultimately leading to the millions of species that have existed. The chemical composition of Earth includes an abundance of hydrogen, oxygen, carbon, and nitrogen forming the basis of millions of distinct organic compounds. Single cell organisms (called prokarya) are still abundant today, with their total weight (biomass) exceeding the biomass of all animal life. Multi-cellular species are called eukarya. All life forms on Earth are formed from complex molecules of deoxyribonucleic acid (DNA). As far as we know, all life on Earth evolved from a single cell. Biologists have developed a system of nine levels (called taxonomy) to classify all life forms. Of the millions of species of life that have existed, more than 99 percent are extinct, primarily due to five major extinction events which accompanied various crises in the history of Earth. Notable among these is the extinction of the dinosaurs 70 million years ago, caused by the impact of a large asteroid. We currently estimate that there are at least eight million species on Earth today, with most undiscovered and undescribed.

Homo Sapiens – Finally, more than 13 billion years after the Big Bang and only two hundred thousand to three hundred thousand years ago, modern humans arrived. One reason it took so long is because we are very complicated creatures, consisting of 7×10^{27} atoms, comprising 40 trillion DNA-based cells. If you could string all of your double-helix DNA end-to-end, it would extend to the Sun and back several hundred times. Every person's DNA is unique and defines their ancestry back to the origin of life on Earth. Each DNA molecule is composed of three billion base pairs which include 46 chromosomes, 23 from each parent. Each chromosome includes sections called genes which define our characteristics. We each have a collection of about 22 thousand genes, called the genome, that define our complete specification. Through evolution, genes are shared among species and continue through time. Our closest relative is the chimpanzee with whom we share 99 percent of our genome. Homo sapiens dominates the planet primarily due to the capability of our brain, the most complex structure known to science. Our brains have permitted us to develop language, collaborate in groups, develop tools and weapons, and to try to comprehend our place in the universe and the world.

Hunter/Gatherers – Our species, Homo sapiens, is the only survivor of about twenty in the family of hominids. The first hominid appeared about three million years ago, and shared many of our characteristics including walking upright and developing primitive tools. Homo sapiens first emerged in Africa between 200 thousand and 300 thousand years ago. We began the road to modern times with little speech, clothing, or weapons and thus were prey as much as predator. Soon human brains and the capability to run for long distances turned us from victim to apex predator. We began to hunt other large mammals to extinction and gathered sustenance from naturally occurring plant life, thus the name "hunter/gatherer." By about 10 thousand years ago, we had populated Earth and had split up into many thousands of small, isolated groups, each with its own language and culture, but with a common desire to understand and control nature through animistic systems of belief. Notwithstanding minor local adaptations (such as skin color), we remain one species with DNA only about one part in a thousand different from any other human.

Civilization – Typically, hunter/gatherer bands were forced to migrate at frequent intervals because of depletion of resources or change in climate. However, in the period from 4000 BCE to 200 BCE, there were four areas that were conducive to permanent residence: Mesopotamia, Egypt, the Indus River Valley, and China. The development of agriculture in each enabled a stable location, rapid growth in population, and development of more complex societies characterized by social stratification, development of writing, and urban growth. Since it was no longer necessary for everyone to grow their own food, other specialties emerged including rulers, tradesmen, artisans, soldiers, and clergy. Each of the four areas made fundamental contributions to future civilizations and invented their own systems of deities. Avenues of trade developed, enabling civilizations to share their accomplishments. The combination of writing, increased population, specialization, and trade accelerated the development of civilizations. Warfare and slavery became the norm. The most successful civilizations conquered large areas and formed empires. Over the six millennia since the birth of Sumeria, there have been approximately 150 distinct civilizations that have risen and fallen with an average lifespan of about 300 to 400 years.

Feudalism to Domination – Following the collapse of the Western Roman Empire in 395 CE, Europe's prospects were less promising than those of Islam or China, which at the time were in their golden ages. The many tiny fiefdoms endured several hundred years of nearly constant threat from external forces, such as the Vikings, and from adjacent kingdoms. The Black Death wiped out 40 percent of the population. However, the Renaissance, originating in Italy, generated new interest in knowledge and learning. The Reformation challenged the authority of the monolithic Catholic Church. Since the Chinese had chosen to trade primarily with India rather than looking eastward, the New World was ripe for colonization by the

Atlantic-facing countries; Portugal, Spain, England, France, and the Netherlands. With improvements in shipbuilding and navigation techniques, the best weapons in the world, and a mercantilist mentality towards extracting wealth from colonies, Europe proceeded to conquer over 80 percent of the world beginning with the voyages of exploration late in the fourteenth century.

Modern Times – We define modern times as the 532 years from Columbus' arrival in the West Indies to the present. In the 300 years following the Columbus voyage, the nations of Western Europe colonized or conquered most of the world due to their advantages in seafaring, weaponry, skill in warfare, and geography, coupled with their immunity to the diseases which annihilated the populations they encountered. Over time, however, the colonists and the conquered developed aspirations for independence, beginning with the American Revolution in 1775. In parallel with their external expansion, the Europeans embarked upon a catastrophic sequence of internal wars, culminating with the two world wars, which left much of Europe exhausted and in ruins. As a result, the modern world partitioned itself into 193 sovereign nations with a wide variety of forms of governance. The Industrial Revolution, starting in England during the nineteenth century, introduced the third great epoch of human civilization (after hunter-gatherers and agriculture), causing the majority of people to move from farms to cities. Modern times can be seen as an on-going struggle between economic systems, notably capitalism vs. communism vs. social welfare, and political systems, in varying forms of democratic republics and autocracies. All of this has occurred within the overall context of continual global growth in scientific and industrial prowess. The United States is still generally perceived as the most powerful nation, but with China growing rapidly, Russia desiring expansion, and North Korea and others as potential rogue states with nuclear arms. The outcomes of modernity are uncertain: Can we avoid large-scale wars; can we preserve the habitability of the planet; will technology be our servant or our master; can we honor the human rights of most people; will democratic republic or autocratic principles prove to the dominant form of governance?

Warfare – For most of us, the outcome of wars fought long ago determined what language we speak, what religion we practice, and whether we are more likely to be affluent or impoverished. Every great civilization has featured a better capacity for warfare than its neighbors and has a willingness, or even thirst, to conquer. The combatant with the most advanced weapons technology and the brightest minds (e.g., Albert Einstein, Leonardo da Vinci, Alan Turing) is likely to win. Technology is always at the service of the military. Every successful civilization must instill in its young males the desire to be warriors, and the evolved tendency of young males to be risk-taking and aggressive supports that need. For the individual soldier, warfare is often a fatal horror, but for his civilization, it is normally a necessity.

Systems of Belief – More than 90 percent of all people adhere to one of five systems of belief: Christianity, Islam, Hinduism, Buddhism, and non-affiliation with religion. In addition, there are thousands of different derivatives and sects with significant followings such as Seventh Day Adventists, Latter Day Saints, Scientologists, Sikhs, Shintoists, and many, many others. Christianity and Islam are strictly monotheistic, one with Jesus Christ as the principal prophet and one with Muhammad. Followers of Hinduism and Buddhism seek enlightenment and rebirth through meditation and good works. The unaffiliated include those who doubt or deny the existence of the supernatural and those who do not know for certain. Globally, there is no majority nor consensus in any system of belief. Therefore, everyone is in the minority.

The Biosphere – We all live together in the tiny volume of Earth called the biosphere which includes the land, air, and water that sustains us. Humankind is placing an enormous load on all three, arising from rapid population growth, increased utilization of energy in all forms, and the continual demand for growth in our standard of living. The greatest attention has been given to global warming, with a growing consensus that if we do not limit the increase to about two degrees Centigrade within a decade or two, the impact on our environment could be catastrophic. Moreover, if the increase ever reaches four degrees Centigrade or more, most or all of the planet will become uninhabitable. Pollution of the land and of the oceans, rivers, and lakes is also approaching intolerable levels. On the land, billions of tons of waste are produced each year overflowing landfills and polluting rivers and lakes. About eight million tons of plastic reaches the oceans annually, killing millions of marine animals. Over 80 percent of the world's sewage is untreated, causing several million deaths per year from disease and chemicals. We depend on the world's forests in many ways, and yet we have destroyed almost two-thirds of the rainforests and are continuing this destruction at the rate of 50 thousand square miles per year. Preserving the habitability of Earth is probably the greatest issue ever presented to humanity, and our long-term survival may depend on overcoming self-interest to enable effective solutions to be achieved.

If you have a moment, please listen to the Woody Guthrie original of the song "Dusty Old Dust" via Google or YouTube. It was written to mourn the climate change in Midwest America in the 1930's Dust Bowl which caused hundreds of thousands of people, like you and me, to lose their homes and their livelihoods. This is what we may all face again in the near future.

Singing: We talked of the end of the world, and then
                         We'd sing a song and then sing it again
                         We'd sit for an hour and not say a word
                         And then these words would be heard

So long it's been good to know yuh,
So long, it's been good to know yuh,
So long, it's been good to know yuh,
This dusty old dust is a-gettin' my home,
And I've got to be driftin' along.

Woody Guthrie
(1912 – 1967)

The Meaning of Life – Ultimately, you may choose for yourself the meaning of your life. It may largely be set for you at birth, at least in terms of your system of belief. In addition, you can add meaning through love of family, a life of service or teaching, study and contemplation, personal achievement, acquisition of wealth and/or power, pleasure, advocacy of a cause, or support of a group, institution, or team. Once you have established the meaning of your life, nothing can deprive you of it, even your death or the death of the universe. It is suggested that you thoughtfully complete the sentence "I AM _____"
to see what it reveals to you. For many, the demands of everyday life and survival may preclude much attention to philosophical matters such as the meaning of life, but even so, they may be leading purposeful, meaningful lives.

The Future – We face the most uncertain century since the last ice age reduced Homo sapiens population to a few thousand. We can imagine a utopian future in which technology and colonization of other planets in other solar systems allow humans to exist far into the future; a dystopian one in which Homo sapiens goes extinct either because we annihilate ourselves, the planet becomes uninhabitable, or Earth is destroyed by natural catastrophe, such as collision with an asteroid. Advancing technologies will impact our lives in the coming century. Some of these include artificial intelligence, nanotechnology, and genetic engineering. Each technology has positive and negative aspects. The Sun will support life on Earth for only one billion more years, but so far, we have existed for only 0.02 percent of that time, so this is not a worry for a very long time.

Ideas – Every important idea has its genesis in a single human mind. Ideas which become influential all traverse a similar path. The originator must be compelling in introducing his idea; followers must be found who will promote the idea; and an infrastructure must develop to provide a home, purpose, and career for believers. If possible, the idea will be institutionalized and belief made a requirement. Finally, adherents will be willing to go to war and die for the idea if opponents arise. The idea's truth is irrelevant to this process, and our species has the characteristic that no idea is too absurd for some of us to believe, given the right promotion. The seven ideas which have had the greatest impact on civilization to date are monotheism, reincarnation, natural rights, capitalism, evolution, Marxism, and the Big Bang. Marxism is an example of a failed idea which remains influential primarily because of its ability to generate hostile reaction. The next great idea has yet to present itself as far as we know.

Conclusion – Prior to the mid-twentieth century, we did not have the capacity to comprehend the creation of our universe or our human evolution and history. The discovery of the structure of the DNA molecule was made in 1953, giving us the ability to understand how all living creatures evolved through time. Background radiation produced by the Big Bang was not verified until 1963. The Hubble Space Telescope, launched in 1990, sent us glorious photos, greatly enhancing our understanding of the beauty and wonder of the trillions of galaxies in the universe. The verification of the Higgs Boson, validating the standard model of the atom, was not accomplished until 2012.

It seems unimaginable today that as recently as the beginning of the twentieth century, experimental verification of the atom had not occurred, the Milky Way was believed to be the only

We are living at the first time in history when science has produced
the tools allowing us to more fully comprehend ourselves, our universe,
and our world.

Although we are totally dependent on Earth as our home, we are in danger
of despoiling it. In that sense, the future of the planet is in our hands.

galaxy, and the understanding of how life on Earth began and evolved was uncertain. Much of today's knowledge did not exist when I was born in 1939. We also know that there is much more yet to be known, including dark matter and dark energy.

Moreover, it is conceivable that we are living in the only place and at the only brief moment of time, over trillions of years, when sentient creatures like you and me can comprehend the universe. The total number of planets is currently estimated at 700 quintillion (that is, 700 thousand billion), so the easy assumption is that there is most likely other intelligent life somewhere out there. But that easy presumption could be wrong! For humans to exist, more than a dozen very low probability conditions had to occur: water in liquid form; a narrow temperature range; a breathable atmosphere; protection from radiation; no catastrophic collisions with asteroids; stable climate; and many others. There are enough of these to make the probability of complex life so low that it may be improbable even over all 700 thousand billion of the other planets.

A confirming study at Oxford University, in 2020, concluded that there is a very low probability of intelligent life anywhere else in the cosmos – a truly surprising, if yet uncertain result. If true, it would resolve the Fermi paradox, "Where is everybody?" with the answer, "There is nobody but us." If so, Homo sapiens may be the only ones ever to comprehend the universe!

So, finally, I suggest again, you and I are two of the luckiest creatures ever to inhabit this universe in its entire 14 billion years, and additionally, two of the luckiest that will ever exist. Because of the magnificent complexity of our brains and the knowledge provided by science, we can vicariously experience everything from the beginning of time to the end of time, and thus understand life forward as well as backward.

I hope you have enjoyed this journey of comprehension and perhaps have learned something new. I contrived to begin the preface of this work with the words of my second cousin, twelve times removed, William Shakespeare and I shall contrive to end it in the same way. After all, William and I share a lot of DNA, as do you and I. We know this with reasonable scientific certainty thanks in part to Charles Darwin and Rosalind Franklin.

———◇◇◇◇———

All's well that ends well.
William Shakespeare
(1564 –1616)

Bibliography

As in the rest of this work, the bibliography makes no claim of completeness. Rather, it is a compendium of books that I found to be particularly useful in doing research and worthwhile recommending that you read.

Actually, in today's world, perhaps the two most efficient research tools are Google and Wikipedia. In using Google, a technique I found particularly helpful was to enter "interesting facts about ..." and this almost always leads to relevant information that would have taken years to dig out of the literature. Similarly, Wikipedia details most topics in a reasonably authoritative and complete way.

Baldwin Richard; The Great Convergence, Information Technology and the New Globalization, Cambridge, MA; Harvard University Press, 2016

Bernstein, William J.; A Splendid Exchange, How Trade Shaped the World, New York, Grove Press, 2008

Brockman, John (Editor); This Explains Everything, New York, Harper Perennial, 2013

Bryson, Bill; The Body, A Guide for Occupants, New York, Anchor Books, 2021

Bryson, Bill; A Short History of Nearly Everything, New York, Broadway Books, 2003

Christian, David; Origin Story: A Big History of Everything; Boston, Little Brown and Company, 2018

Cox, Brian; Wonders of the Solar System, New York, Harper Collins Publishers, 2010

Davidson, Peter; Atlas of Empires, The World's Great Powers from Ancient Times to Today, Mount Joy, PA, Fox Chapel Publishing, 2018

Dawkins, Richard; The Magic of Reality, New York, 2011

Diamond, Jared; Guns, Germs, and Steel, The Fates of Human Societies, New York, W.W. Norton and Company, 1997

Feynman, Richard P.; Six Easy Pieces, New York, Basic Books, 1963

Friend, David (Editor); More Reflections on the Meaning of Life, Boston, MA., Little Brown and Company, 1992

Golub, Leon and Pasachoff, Jay M.; Nearest Star, The Surprising Science of Our Sun, Cambridge, MA., Harvard University Press, 2001

Greene, Brian; Until the End of Time, New York, Vintage Books, 2020

Grun, Bernard; The Timetables of History, New York, Simon and Schuster, 1975

Harari, Yuval Noah; Sapiens, A Brief History of Humankind; New York, Harper Perennial; 2015

Harari, Yuval Noah, Homo Deus, A Brief History of Tomorrow, Harper Perennial, 2017

Harari, Yuval Noah; 21 Lessons for the 21st Century, New York, Spiegel and Grau, 2018

Hawking, Stephen W.; The Theory of Everything, The Origin and Fate of the Universe, Beverly Hills, CA, New Millennium Press, 2002

Keegan, John; A History of Warfare; New York, Vintage Books, 1994

Krauss, Lawrence M.; A Universe from Nothing, New York, Atria Paperback, 2012

Lanza, Robert; Biocentrism; Dallas, TX; BenBella Books, 2009

Leeming, David with Leeming, Margaret; A Dictionary of Creation Myths, New York, Oxford University Press, 1994

Parsons, Paul; The Beginning and End of Everything, From The Big Bang to the End of the Universe, London, Michael O'Mara Books Limited, 2018

Reader's Digest (Editors); The Last Two Million Years, Pleasantville, NY, The Reader's Digest Association, Inc., 1977

Sagan, Carl; Cosmos, New York, Random House, 1980

Wilkerson, Isabel; Caste, The Origins of Our Discontents; New York, Random House, 2020

Acknowledgements

I can no other make but thanks, and thanks, and ever thanks.
William Shakespeare
(1564 – 1616)

I would maintain that thanks are the highest form of thought,
and that gratitude is happiness doubled by wonder.
G. K. Chesterton
(1874 – 1936)

My wife, Linda, my son Greg Martin, and my son-in-law Martin Lewis reviewed every word. In addition to making many useful comments and corrections, they protected me from poor writing habits and from making too many lame attempts at humor. My professional colleague, Douglas Raymond, corrected several subtle, but egregious, errors of fact. My daughter, Shelley Martin, made the useful suggestion that I retain a professional editor. Marti Childs of EditPros LLC in Davis, CA, served admirably in that role.

In a work of this scope, full of dates, concepts, ideas, numbers, estimates, and even opinions, errors have almost certainly crept in, and I apologize for them. They are all my fault. Should you find any, I would be pleased to learn of them at **sumandsub39@ gmail.com** so that they might be corrected. Your comments on the content are welcome as well. To the extent that you might disagree, you may tell me if you wish, but you will hear no rebuttal from me since this book is about comprehension, not persuasion.

About the Author

———∞∞∞———

Every man has a book in them, but in most cases that's
where it should stay.
Christopher Hitchens
(1949 – 2011)

Another damned thick heavy book! Always scribble,
scribble, scribble, eh, Mr. Gibbon?
Duke of Gloucester
(1743 – 1805)

Bill Martin has Bachelor's and Master's Degrees in Electrical Engineering from Stanford University (1961 and 1962) and, in 1975, a Master's in Management from the Massachusetts Institute of Technology (MIT). He began his career in the defense industry as a designer of computers and military command and control systems. He led a major commercial Air Traffic Control program and was principal investigator of a NASA-sponsored study of ultra-reliable computing. As a participant in highly classified programs, he gave senior-level briefings at the Pentagon and NATO headquarters. In the Sloan School of Management at MIT, he created a comprehensive simulation model of the future of health care. He taught a course in advanced communications statistics at the University of Southern California (USC) and for twenty years taught college-level mathematics with highest ratings from his students. Later, he was vice-president of engineering for a Northern California test equipment firm. He managed the development of a family of circuit board testers with over $100 million in sales. In retirement, he tutored local students in mathematics. He was elected to public office and served as trustee at Sierra College, Placer County, California. He attended six Rose Bowls and one Final Four. Bill lives in Northern California with Linda, his wife of 62 years. His two children and their spouses live nearby.

Made in the USA
Middletown, DE
23 December 2022